Cloud and Serverless Computing for Scientists

Juan A. Añel • Diego P. Montes
Javier Rodeiro Iglesias

Cloud and Serverless Computing for Scientists

A Primer

Juan A. Añel
Universidade de Vigo
Ourense, Spain

Diego P. Montes
Universidade de Vigo
Ourense, Spain

Javier Rodeiro Iglesias
Universidade de Vigo
Ourense, Spain

ISBN 978-3-030-41783-3   ISBN 978-3-030-41784-0   (eBook)
https://doi.org/10.1007/978-3-030-41784-0

© Springer Nature Switzerland AG 2020
This work is subject to copyright. All rights are reserved by the Publisher, whether the whole or part of the material is concerned, specifically the rights of translation, reprinting, reuse of illustrations, recitation, broadcasting, reproduction on microfilms or in any other physical way, and transmission or information storage and retrieval, electronic adaptation, computer software, or by similar or dissimilar methodology now known or hereafter developed.
The use of general descriptive names, registered names, trademarks, service marks, etc. in this publication does not imply, even in the absence of a specific statement, that such names are exempt from the relevant protective laws and regulations and therefore free for general use.
The publisher, the authors and the editors are safe to assume that the advice and information in this book are believed to be true and accurate at the date of publication. Neither the publisher nor the authors or the editors give a warranty, expressed or implied, with respect to the material contained herein or for any errors or omissions that may have been made. The publisher remains neutral with regard to jurisdictional claims in published maps and institutional affiliations.

This Springer imprint is published by the registered company Springer Nature Switzerland AG.
The registered company address is: Gewerbestrasse 11, 6330 Cham, Switzerland

Foreword

As a researcher in a number of different domains over my career, all of which have benefited from the development in the areas of distributed computing, it is clear that we will continue on this journey of discovery and development of new tools and services for some time to come. We moved away from historic paradigms, such as the mainframe, with the rise of the personal computer and from then onto the increasingly powerful workstation, generic high-performance computing, grid computing, and now "the cloud." One disadvantage of these changes though is that proponents of these new architectures often sell the newest paradigm as being better in nearly all ways than those that have proceeded it and therefore it should replace the systems following the older paradigms, moving users onto newer, bigger, better shinier systems. This is plainly not the case once hype has been ignored and as such the content of this book should be considered as aiming to educate the reader in terms of making sure that the right tool is always used for the right job.

I was struck when reading the book by the diversity of target audience within the research field, making sure that communities of practice, where the use of cloud will become ever more valuable are covered, with relevant examples is a great feature. I would though also make sure that one further cohort reads this book, the project PI or senior professor for whom this book should be required reading. Alongside more traditional responsibilities they are also now the ones who are under legal obligations to comply with requirements on open data and reproducibility, something that the migration of services and data storage to the cloud is extremely suitable for. They should also ensure that they understand the benefits and almost more importantly the challenges that a cloud-first strategy can bring.

When working with Juan and Diego prior to the development of this book, we spent a significant amount of time working through different models of the utilization of Infrastructure as a Service cloud computing for the environmental sciences, in particular climate science. Therefore, a large number of points made by this book have been developed through practical experience of the authors, not through desk-based research. Along with this with my background in the development of pan European Federated Cloud services for research, I can clearly see how lessons learnt, sometimes in very hard ways, in how researchers should be thinking about using the cloud could have been found quicker and more easily if such a text as this existed, or that we could have shared with members of the user community. As such I also see this book being of great use for those that will in time become users not only of large-scale public clouds to support their research but also in the usage of the European Open Science Cloud as it becomes a more and more useful tool to European Research.

I would finally highlight how the authors already say that in the fast-moving and dynamic world of cloud computing that things get out of date very quickly. Examples of this of course abound, though a key highlight is where the different deployment models of cloud computing mean that it is not always the same cloud that two different researchers talk about when they describe their method of use of "cloud." Therefore, making sure that everyone understands the underlying model of cloud computing as is done early within the book is essential. So to conclude I would say that readers should, once they think they understand how they can use cloud computing within their research, stop, read, and then think about how they want to move forward with the cloud, making sure that they follow sound scientific and IT principles in how they deploy services within the cloud, testing all the time and recognize that cloud is not a magic bullet, but in the longer term it can make things much, much easier if their deployments are well designed.

Associate Professor and Associate Prof David Wallom
Director – Innovation Oxford Research
Centre University of Oxford
Oxford, UK

Preface

We have written this book for scientists, engineers, or anyone who wants to approach cloud computing, simply to know more about it or if they are considering or evaluating it as an alternative or complementary solution for their computing needs. Additionally, this book can be useful for IT professionals, such as Solutions Architects, who want to learn about the current and future needs of the scientific community and how they can be satisfied using cloud technologies, enabling them to offer more suitable proposals to their customers. Therefore, this text can provide a bridge between user and provider perspectives.

Moreover, this book attempts to provide a broad view of the current status of cloud computing and to serve as an introductory text for graduate, postgraduate, and Ph.D. students across every discipline, providing insight into potential uses and how it can help to develop projects and future work.

Ourense, Spain	Juan A. Añel
Ourense, Spain	Diego P. Montes
Ourense, Spain	Javier Rodeiro Iglesias
November 2019	

Acknowledgements

We want to dedicate the book to our family and friends for their daily support and understanding, which makes it possible for us to devote our time to our research work, too often at the expense of our time with them.

We would also like to thank Springer for the opportunity to publish this book and their interest in it, especially Zachary Romano, our publishing editor here, and the AJE language editing service. Special thanks to Paulo Rodríguez from Dropbox for facilitating part of the work that has contributed to this book and useful discussions. Additionally, we would like to thank Google Enterprise, Microsoft Research, and Amazon Web Services for granting us computational resources and facilitating technical support to perform experiments using their cloud computing platforms. We are deeply grateful to Tom Grey, Barak Regev, and Shailesh Rao from Google and Kenji Takeda from Microsoft Research. We are also grateful to our colleagues at ClimatePrediction.net at the University of Oxford, in particular David Wallom, all the staff from the Supercomputing Centre of Galicia (CESGA), in particular Carlos Fernández, and Tomás Fernández Pena of the Universidade de Santiago de Compostela.

We thank Dr. Sulay Tovar from the CIMUS (Universidade de Santiago de Compostela) and Prof. José Agustín García from the Universidad de Extremadura for providing feedback on draft versions of this book. Also, we thank Alexandra Iglesias for her contribution to the design of the cover of the book.

Part of the research that we have developed for years, therefore enabling us to accumulate the knowledge to write this book, has been funded by different agencies and bodies, among them is the contribution by the European Regional Development Fund (ERDF). Additionally, Juan A. Añel has been supported in recent years with a Ramón y Cajal Grant from the Government of Spain (RYC-2013-14560).

Contents

1	**Why This Book?**		1
	References		3
2	**Why the Cloud?**		5
	2.1	Introduction	5
	2.2	Requirements and Needs	6
		2.2.1 Sasha in Their House or with a Self-employed Business	7
		2.2.2 Sasha as Entrepreneur	7
		2.2.3 Sasha in Their Laboratory	10
		2.2.4 Sasha Has an Idea	13
		2.2.5 Sasha Needs Mobility	14
	2.3	Everyone Is Sasha	14
	References		15
3	**From the Beginning to the Future**		17
	3.1	What Is Cloud Computing?	17
	3.2	Types of Cloud	18
		3.2.1 Commercial and Non-commercial	18
		3.2.2 Cloud Models	20
		3.2.3 Types of Service	21
	3.3	Features of the Cloud	23
		3.3.1 Elasticity and Scalability	23
		3.3.2 Self-service Provisioning	23
		3.3.3 Application Programming Interfaces	24
		3.3.4 Performance Monitoring and Measuring	24
	3.4	Cloud Native Concepts	24
		3.4.1 Cloud Native	25
		3.4.2 Containers	25
		3.4.3 Microservices	26
		3.4.4 Kubernetes	26
	3.5	Environmental Issues	26

3.6	Success Stories of Cloud and Serverless Computing in Science		27
	3.6.1	Archeology	27
	3.6.2	Computer Sciences	28
	3.6.3	Energy	28
	3.6.4	Geosciences	28
	3.6.5	Library Science	28
	3.6.6	Mathematics	29
	3.6.7	Medical Sciences	29
	3.6.8	Physics	30
	3.6.9	Social Sciences	30
	3.6.10	Space Research	30
References			30

4 Show Me the Money 33
- 4.1 Money for Research Using Cloud Computing 33
- 4.2 Sharing Costs 34
- 4.3 Selling Research 35
- 4.4 What Happens If I Do Not Pay? 35
- 4.5 What Happens If My Provider Does Not Deliver 36
- 4.6 Assessing Migration 36
  - 4.6.1 Selecting a Cloud Computing Provider 37
  - 4.6.2 Problem Definition 37
  - 4.6.3 Gather Requirements 37
  - 4.6.4 Obtain the Data 39
  - 4.6.5 Review and Decide 39
- 4.7 Conclusions 40
- References 40

5 Tools in the Cloud 41
- 5.1 Sasha Wants Better Tooling for Their Organization 41
- 5.2 General Productivity 41
  - 5.2.1 Office 42
  - 5.2.2 Communication 42
- 5.3 Scientific and Engineering 43
  - 5.3.1 Document and Paper Edition and Sharing (LaTeX) 43
  - 5.3.2 Project and Issue Management 43
  - 5.3.3 Version Control 44
  - 5.3.4 Other Relevant SaaS 45
- 5.4 Sasha Decision 45
- References 46

6 Experiments in the Cloud 47
- 6.1 Sasha Is Ready to Start a New Experiment 47
- 6.2 Running Code on the Cloud 48
  - 6.2.1 Understanding Software Development on the Cloud 48
  - 6.2.2 Application Programming Interfaces (APIs) 49

		6.2.3	Cloud Environment and Runtimes	50
		6.2.4	Code in the Cloud: Considerations	53
	6.3	Starting an Experiment in the Cloud		55
		6.3.1	Example: Month Classifying Experiment	55
	References			61
7	**Serverless Experiments in the Cloud**			63
	7.1	Sasha Revisits the Experiment		63
	7.2	Thinking Serverless		64
		7.2.1	Pros and Cons	64
	7.3	Example: Month Classifier, Serverless Version.......................		65
	7.4	Example: Serverless Sensor Data Collector		70
	References			72
8	**Ethical and Legal Considerations of Cloud Computing**			73
	References			75
9	**You Are Outdated, We Are Already Updating This Book**................			77
Glossary				79
Index				83

Acronyms

AI	Artificial Intelligence
API	Application programming interface
AWS	Amazon Web Services
BOINC	Berkeley Open Infrastructure for Network Computing
CDC	USA Centers for Disease Control and Prevention
CERN	European Organization for Nuclear Research—*Conseil Européen pour la Recherche Nucléaire*
CLI	Command Line Interface
CNCF	Cloud Native Computing Foundation
DoD	US Department of Defense
EC2	(AWS) Elastic Cloud Compute
ERP	Enterprise Resource Planning
ESA	European Space Agency
EU	European Union
FaaS	Function as a Service
FDA	U.S. Food and Drug Administration
FMI	Finnish Meteorological Institute
GCP	Google Cloud Platform
GCS	Google Cloud Storage
GPL	GNU Public License
HIPAA	Health Insurance Portability and Accountability Act
HIV	Human Immunodeficiency Virus
HPC	High performance computing
HPCaaS	High performance computing as a Service
IaaS	Infrastructure as a Service
IOPS	Input/output operations per second
IoT	Internet of Things
IRC	Internet relay chat
IT	Information Technologies
JPL	Jet Propulsion Laboratory
JSON	JavaScript object notation

Met Office	United Kingdom's National Weather Service
MPI	Message Passing Interface
NAS	Network Attached Storage
NASA	National Aeronautics and Space Administration
NOAA	USA National Oceanic and Atmospheric Administration
NREL	National Renewable Energy Laboratory
OS	Operating system
PaaS	Platform as a Service
POSIX	Portable Operating System Interface
PUE	Power Usage Effectiveness
RAM	Random-Access Memory
S3	AWS Simple Storage Service
SaaS	Software as a Service
SDK	Software Development Kit
SEO	Search Engine Optimization
SSD	Solid-State Drive
SSL	Secure Sockets Layer
TLS	Transport layer security
UI	User Interface
VPC	Virtual private cloud
VPN	Virtual private network
VPS	Virtual private server

Chapter 1
Why This Book?

Since the advent of computers, scientific research has intensively used and has continuously pushed their development. Computers are used for research in very different ways, with high-performance computing (HPC) and data storage two of the most demanding uses. Such uses have always required massive on-site hardware; however, with the advances in the cloud computing paradigm over the past decade, and more recently serverless computing, this requirement no longer exists. Indeed a top500.org report [1] shows that for the period 2016–2017, the HPC market grew by approximately 1.6%; meanwhile, for the same period, use of cloud HPC services increased from 7 to 44%. The rationale for this difference appears to be that HPC as a Service (HPCaaS) is being used to complement resources from on-site HPC infrastructures when such resources are not sufficient to satisfy demand. In fact, this is one of the most appealing features driving the expansion of cloud computing in research: to satisfy the demand for computing resources for punctual projects.

Cloud computing has been a mainstream technology for over 10 years and is widely used and very well known in IT departments. The applications, solutions, and implementations of cloud computing are being widely explored. However, it is not so common that users are fully aware of its potential, uses, internals, or market conditions. That is, people have become familiar with the main characteristics of a computer, laptop, workstation, etc., and have basic knowledge on what they can do with such technology and how to choose one to purchase according to memory or processors; however, this is not yet the case for cloud computing. Many people have not adapted to this new paradigm, driving a knowledge gap between the currently prevalent technology and the technology familiar to users. Regardless, a large percentage of the population is using cloud computing for daily work or for leisure via an application on a smartphone. This is also true among those working in scientific disciplines, conducting research, or acting as lab leaders, research center directors, or decision makers. These people could obtain many advantages by at least partially adopting cloud computing, as will be demonstrated later in this book through case studies on its application. This applies to almost every discipline of science. It is also the case for decision makers who might have to consider transitioning IT infrastructure to cloud computing and want to know more about the implications. A report from 2016 by The Economist Intelligence Unit discussed how lack of knowledge and trust are barriers for the adoption of cloud computing [2]. Although these numbers have probably improved, a recent survey by Eurostat found that only 28% of businesses in Europe use cloud computing and that broader adoption is expected in the future [3].

A more recent concept is serverless computing, which is in some ways an evolution of cloud computing, that attempts to increase the level of abstraction for users so that they need to be less concerned about the computational infrastructure they use.

From the perspective of research funding bodies, both the public and private have studied and applied cloud computing solutions for the infrastructure and projects that they fund. This can be achieved in various ways, for example, by deploying cloud infrastructure using publicly owned HPC centers or granting vouchers to use cloud computing resources from commercial providers instead of granting computing time in public HPC facilities.

Another idea behind the broader use and adoption of cloud computing is that it presents great possibilities to democratize the development of scientific research, as in-house HPC facilities no longer must be built and maintained. This advantage is especially important for funding bodies, researchers, and laboratories that do not have access to such facilities and those that have obtained funding for a specific time-boxed project and do not want or need to invest the money to buy and maintain hardware that will not be used in the future or will become deprecated. This development has great potential as it eliminates barriers to research in the most disadvantaged countries and regions of the planet, that is, those in which the R&D infrastructure is underdeveloped.

A very important issue is the fact that cloud computing has the potential to be combined with other technologies such as big data and machine learning to enhance scientific productivity and research advancement. Integration with these technologies takes advantage of the potential combination of features such as data storage or running models in cloud systems. Such integration enables direct access to complete solutions already implemented by the vendor and adapted entirely to the cloud computing environment, thereby removing part of the technical burden and potential problems when adapting or configuring services. This process makes it straightforward to perform data mining analysis on research data stored in the cloud, creating a great potential to obtain new and unexpected results and advances. When these techniques are combined effectively, the approach is called "cloud data mining" [4].

Another benefit of the combination of all these technologies is that it provides a way to achieve a "zero download" scenario, solving one of the existing problems in making research data available: an excessive number and size of download requests that drives the need for massive local storage resources. For some scientists, laboratories, and fields of research, this approach represents an almost complete cultural change, from downloading data to compute results locally to performing computations on data in the cloud and downloading only the final results.

Finally, from the perspective of the job market, job offers in the field of lead or director of scientific computing reflect the existing movement to cloud computing, with specific requests that applicants in these roles have knowledge to manage private and public cloud infrastructure. At this point, it is necessary for those working in fields of science and research to obtain a quick and complete view of the ecosystem of cloud computing, the challenges and opportunities that it offers to enhance scientific research, the efficiency in expenditure and environmental footprint.

Therefore, this book is intended to address these issues, to fill the gaps here described, and to provide elements to better assess what benefits can be achieved by migrating to cloud computing. If you are a research manager heading a research center, a scientist using computing resources with superficial or no knowledge on cloud computing and serverless technologies, a member of the IT department of a laboratory or a seller of IT technologies, we hope that this book provides a brief and useful tool.

References

1. Feldman M (2018) Cloud computing in HPC surges. https://www.top500.org/news/cloud-computing-in-hpc-surges/. Accessed 29 Nov 2019
2. The Economist Intelligence Unit (2016) Trust in cloud technology and business performance: reaping benefits from the cloud, 26 pp.
3. Kaminska M, Smihily M (2018) Cloud computing – statistics on the use by enterprises. https://ec.europa.eu/eurostat/statistics-explained/index.php?title=Cloud_computing_-_statistics_on_the_use_by_enterprises&oldid=416727. Accessed 9 Apr 2019
4. Barua HB, Mondal KC (2019) A comprehensive survey on cloud data mining (CDM) frameworks and algorithms. ACM Comput Surv 52(5):104. https://doi.org/10.1145/3349265

Chapter 2
Why the Cloud?

2.1 Introduction

The reasons for using cloud computing resources differ among users. In some ways, the use of cloud computing depends on whether the intended use is for high-performance computing (HPCaaS), data storage or simply to host responsive web services (Infrastructure as a Service (IaaS)) or to perform less demanding tasks such as organizing research activity (email, code, laboratory task organization systems, etc.). In all these cases, cloud computing can provide advantages such as the possibility of needing smaller and less powerful computers and lower costs in terms of premises for daily laboratory work. Using a service provided by a vendor also has advantages in that you or the IT staff can focus on working on advancing research projects (coding, improving computational efficiency of your daily work, etc.) instead of spending the time on solving hardware maintenance problems.

Vendors promote the use of cloud computing for research, claiming advantages such as quick access to resources (that is, cloud computing provides access in minutes to something that could take weeks: to order new hardware, receive it, install software and run your experiments), lower costs (as you pay only for what you actually use instead of having your computers idle), and flexibility in the resources that you contract through adaptation of your project to the amount of money that you have available. An example of the last point is running your experiments only when the cost of doing so is below a threshold that you have previously established or, when you are in a rush because of a hard deadline, having the possibility of running unlimited experiments simultaneously and choosing computers with greater performance. An additional point is computer security, which is usually claimed by cloud computing vendors to be of the highest level and for most laboratories is probably better than what they can achieve with in-house staff, as having IT staff specialized in security is not standard.

One more point that supports the adoption of cloud computing in research is enhancing scientific collaboration, that is, facilitating collaborative research and innovation. If you have tools, models, or data that a colleague in other part of the world wants to use, you do not need to send anything and wait for long data transfers, etc. A good use of cloud computing services would enable you to simply share a copy, and the other colleague would cover any cost associated with their work.

That is, you can stop caring about payments for physical support for the data (drives, etc.), delivery fees or border controls and taxes.

Another issue is the potential time saving that cloud computing offers to the final user who does not need to be concerned about software updates, compatibility, or security patches. All this information is transparent to you, with the advantage, for example, of having available at any moment several versions of the same software so you can use the one that best fits to your needs for each case.

2.2 Requirements and Needs

This chapter is oriented to a non-technological user and tries to show typical examples or common strategies that such a user can find via an Internet search on cloud computing. We will assume the main elements necessary for working on the cloud, such as having a good Internet connection with the necessary bandwidth and having a personal computer with sufficient power to run all the services that may be required from the cloud. These two elements are mandatory by definition to access services and products on remote servers or infrastructure.

2.2 Requirements and Needs

For these examples, Sasha is a person that we could meet anywhere in the morning while walking in the spring sun. Sasha could have various occupations, and we are going to attempt to illustrate how cloud computing can help Sasha to achieve the goals of these occupations. The specific scenarios considered are as follows:

- Sasha has a small business or is self-employed and works from home.
- Sasha has a business with employers and has many costumers or providers who need to perform multiples operations a day.
- Sasha manages a research laboratory and works with other researchers.
- Sasha has an idea for a service that could bill to costumers.
- Sasha needs mobility.

2.2.1 Sasha in Their House or with a Self-employed Business

Let us assume that you are Sasha, a professional who works from home. In your daily work, you develop computer applications for others, write source code and technical records for clients, manage many documents and draft versions, and have frequent conversations with providers and clients.

Sasha could build a system for work at home: a fiber Internet connection, a server, a storage service through NAS, a Git server to maintain projects and draft versions, a tool for videoconferencing, programming backup, etc. However, Sasha only has time for their job and cannot manage and support the system. Additionally, Sasha currently does not have the money to purchase all the technology necessary to fulfill the needs described. Moreover, Sasha is not an IT professional and simply wants to focus on doing the job (Tables 2.1, 2.2, 2.3, and 2.4).

If Sasha does not want to pay money to obtain all this infrastructure, they could try to obtain free tools on the cloud. Clearly, Sasha is a professional and would like to have all the technological possibilities for work, but the free options are unlikely to provide all these abilities.

Let us consider the alternatives for each of the technologies that Sasha needs.[1]

2.2.2 Sasha as Entrepreneur

Sasha has a business with employees, a business where it is necessary to interact with customers and suppliers. In this business, Sasha needs to issue invoices, make budgets, and exchange information with customers and suppliers at the shipping and transportation levels. Additionally, Sasha needs information about accounting and logistics to track shipping to customers.

---

[1] The technologies discussed are only a part of all the existing options, and the estimated costs shown are from October 2019 and can change depending on the country of the client.

Table 2.1 Comparison of costs for several cloud storage options

Cloud storage				
Tool	Free	Cost	Paid	Cost
Google drive cloud storage [1]	15 GB	0	100 GB	1.99 $/month
IDrive [2]	5 GB	0	2 TB	69.5 $/year
Mega cloud storage [3]	50 GB	0	200 GB	12 €/month
One drive cloud [4]	5 GB	0	100 GB	1.99 $/month
Box cloud storage [5]	10 GB	0	100 GB	9 €/month
ICloud cloud storage [6]	5 GB	0	50 GB	0.99 $/month

Table 2.2 Comparison of costs for several Cloud software repository and version control options

Cloud software repository		
Tool	Cost	Description
GitHub Free [7]	0 $ user/month	Unlimited public and private repositories. Basic functionalities
GitHub Pro [7]	7 $ user/month	Unlimited public and private repositories. Pro tools for developers with advanced requirements
GitHub Team [7]	9 $ user/month	Unlimited public and private repositories. Advanced collaboration and management tools for teams
GitLab Free [8]	0 $ user/month	Unlimited public and private repositories. Helping developers build, deploy, and run their applications
Gitlab Bronze [8]	4 $ user/month	Unlimited public and private repositories. Enabling teams to speed DevOps delivery with automation, prioritization, and workflow
Gitlab Silver [8]	19 $ user/month	Unlimited public and private repositories. Enabling IT to scale DevOps delivery with progressive deployment, advanced configuration, and consistent standards
Gitlab Gold [8]	99 $ user/month	Unlimited public and private repositories. Enabling businesses to transform IT by optimizing and accelerating delivery while managing priorities, security, risk, and compliance
Bitbucket Free [9]	0 $ user/month	Unlimited public and private repositories. Free for up to 5 users
Bitbucket Standard [9]	3 $ user/month	Unlimited public and private repositories. For growing teams who need more
Bitbucket Premium [9]	6 $ user/month	Unlimited public and private repositories. For large teams with advanced features

Sasha needs information systems that enable storage and management of all the information related to the transactions with customers and suppliers. These systems must be able to offer customers a secure, reliable, and usable platform to make their purchases and to track the status of their shipments. Moreover, the systems must provide Sasha with up-to-date information on how each sale is made and about the habits and preferences of customers in order to offer new products. For suppliers,

Table 2.3 Comparison of costs for several cloud video conferencing options

Cloud video conferencing				
Tool	Free	Cost	Paid	Cost
Skype [10]	Personal	$0	Business	Included in Office 360
Skype [10]	Personal	$0	Business	$159.75 (Included in Office 360)
GoToMeeting [11]	–	–	Personal	$20.94/month
Zoom [12]	Basis	$0	Enterprise	$20.92/month (host)
Webex [13]	Libre	$0	Starter	$14.16/month

Table 2.4 Comparison of costs for several cloud backup options

Cloud backup		
Tool	Cost	Description
Backblaze [14]	6 $/month/computer	Unlimited data
Carbonite (one computer) [15]	6 $/month	Unlimited data
Carbonite (multiple computers) [15]	24 $/month	Unlimited data
Carbonite (computers + servers) [15]	50 $/month	Unlimited data
CrashPlan [16]	10 $/month/computer	Unlimited data
Livedrive Backup (one computer) [17]	6.61 $/month	Unlimited data
Livedrive Pro Suite (five computer) [17]	18.73 $/month	Unlimited data
IDrive Basic [18]	Free	5 GB
IDrive Personal [18]	69.50 $/year	2 TB
IDrive Business [18]	99.50 $/year	250 TB

the information system must be able to follow the status of orders, manage product inventory, and provide the necessary accounting in terms of economic capacity to make the corresponding payments.

Sasha has several possibilities to obtain these services. On the one hand, it is possible to develop an infrastructure on their premises to install the information systems. However, Sasha does not want to purchase, configure, or manage their own servers. Another option is to hire servers in the cloud and install the services on them. However, as in the previous case, Sasha does not want to configure or manage the services. The last option is to hire the information service directly (as an ERP or e-commerce builder). Sasha selects this final option. Therefore, the only requirement is to enter the information into the system and provide access to customers and suppliers so that they can work directly in their information system. With this option, all Sasha's business information is in the Cloud, where it could benefit from further advantages (if hired) such as backups or redundant systems that guarantee the availability of the service permanently or technical service to solve problems 24 h a day.

Sasha can decide between several possibilities when hiring an e-commerce builder depending on the number of functionalities and size. Of course, the cost will increase with the scope. The implementation of the information system for the company still needs to be able to enter the market with a suitable image, a set of

Table 2.5 Comparison of costs to build a website

Estimated cost of building a website [19]		
Factors	Hiring a web designer	Using a website builder
Setup	$160	$0
Design and building	$5000	$0
Content creation	$50	$0
Training to use it	$60	$0
Maintenance	$500	$60
Total	$6760	$60

Table 2.6 Example of e-commerce website pricing [20]

	Small	Mid-size	Enterprise
Design	$5000	$10,000	$35,000
Programming	$2000	$15,000	$75,000
Integrations	$500	$8000	$20,000
Data import	$0	$5000	$10,000
Hosting (Annual)	$500	$6000	$10,000
SEO (Annual)	$12,000	$36,000	$60,000
Average e-commerce website cost (Annual)	$20,000	$80,000	$210,000

elements, including a corporate design, programming to satisfy specific functions of the business, integration of functionalities and data import, and to hire a hosting service that includes a domain and implementation of SEO techniques for better positioning in Internet search engines.

Tables 2.5 and 2.6 show cost estimates for deploying the discussed infrastructure.

2.2.3 Sasha in Their Laboratory

Let us assume now that Sasha is a researcher in a research center that has a budget to cover the needs of the laboratories. We assume that the research that Sasha conducts produces considerable amounts of information that must be processed to extract knowledge. Sasha therefore needs computers on which to process information and data storage to have the information available at any time. We assume, as in the previous two cases, that Sasha's laboratory has an Internet connection and sufficient bandwidth to access services outside the laboratory. Sasha's field of knowledge is not computer science and Sasha cannot, and does not want, to create the infrastructure to conduct research on the data. The problem of data storage is considered in detail in the first case discussed in this section; the need for processing is now considered.

Sasha could hire a virtual server, dedicated or not, to perform data analysis: this solution can be implemented at several levels. At the lowest level, Sasha can hire a virtual server without any pre-configuration, but this solution requires Sasha to configure almost everything on the virtual server (this scenario is called IaaS). On a higher level, Sasha can hire a virtual server that has a partially pre-configured system, which requires Sasha to manage and configure only a portion of the system (Platform as a Service (PaaS)). The best option for Sasha is a virtual server with a fully configured system that is readily hired (SaaS).[2] In this situation, Sasha simply has to prepare and execute the data analysis process.

This hired virtual server has several features that can vary the cost:

- **The number of cores:** Determines the number of simultaneous processes that can be executed on the server.
- **The type and storage capacity:** Higher storage access speed and greater storage capacity increase the cost of the server.
- **Dedicated RAM memory of the server:** If the virtual server has more RAM memory, less time is required to access the disk to retrieve information and processing is faster.
- **Virtual server bandwidth:** Higher bandwidth means less transfer time between Sasha's laboratory and the virtual server in the cloud (the transfer time is also limited by the bandwidth Sasha has in the laboratory).
- **The type of support:** Support provided 24/7 is more expensive than less complete support.
- **A digital certificate (TLS):** Having a digital (TLS) certificate that guarantees the protection of the information between the virtual server and all the computers that communicate with it increases the cost.

As indicated above, when selecting a virtual server, Sasha can choose between a managed or self-managed server. A self-managed virtual server is one that must be installed and configured by the user (remember, Sasha does not want to do it). A self-managed virtual server is one that incorporates utilities that make its use easier for non-technical users. A self-managed server can provide tools such as the following:

- automatic fast software installation
- server management and security updates
- email, database, and domain settings through a control panel
- free private SSL certificate and automatic backups

Table 2.7 provides some examples and comparisons of costs for existing solutions in the market.

---

[2]IaaS, PaaS, and SaaS are types of cloud computing services that are discussed in detail in Sect. 3.2.3.

Table 2.7 Comparison of costs for several virtual private servers

Some examples of VPS [21]

VPS	Cores	SSD	RAM	Bandwidth	Free domains	Support	Free SSL/TLS	Cpanel	Cost/Month
Bluehost	2	30 GB	2 GB	1 TB	1	24/7	1	–	$18.99
Bluehost	2	60 GB	4 GB	2 TB	2	24/7	1	–	$29.99
Bluehost	4	120 GB	8 GB	3 TB	2	24/7	1	–	$59.99
HostGator	2	120 GB	2 GB	1.5 TB	–	24/7	–	–	$29.95
HostGator	2	165 GB	4 GB	2 TB	–	24/7	–	–	$39.95
HostGator	4	240 GB	8 GB	3 TB	–	24/7	–	–	$49.95
iPage	1	40 GB SSD?	1 GB	1 TB	1	24/7	–	Yes	$19.99
iPage	2	90 GB SSD?	4 GB	3 TB	1	24/7	–	Yes	$47.99
iPage	4	120 GB SSD?	8 GB	4 TB	1	24/7	–	Yes	$79.99
JustHost	2	40 GB SSD	2 GB	5 TB	–	24/7	–	Yes	$69
JustHost	8	150 GB SSD	8 GB	5 TB	–	24/7	–	Yes	$139
JustHost	16	200 GB SSD	16 GB	5 TB	–	24/7	–	Yes	$189

Question marks indicate that the storage is assumed to be SSD, but it is not specified by the vendor

2.2.4 Sasha Has an Idea

Sasha has an idea to improve the bureaucratic processes in their laboratory and to increase the impact of their research output. Sasha wants to automate handwritten orders for contracts with funders and orders of materials. To do this, Sasha has developed a service called *MyBillProcess*. With this service, a registered user can take a photo of a handwritten order and send it electronically to the service from anywhere on the planet. The novelty of the *MyBillProcess* service is that the service scans the image of the order and recognizes the details of the client, stakeholder, or provider and the products and units of each product to be ordered. A user registered with the *MyBillProcess* service has access to the laboratory catalogue including services, research outputs, examples, and success stories, with all the necessary information, including stocks of materials and needs. The *MyBillProcess* service uses the recognized order data to create a digital order with the data of the products that the registered user has in the system and sends this digital order to the office, including finances or the storehouse to be processed, or to the appropriate person or division in the laboratory. The *MyBillProcess* service uses the scanned data and data from others to build profiles and potential offers or newsletters for stakeholders that could be sent by email to increase the engagement with stakeholders and to promote the research output of the laboratory.

As in previous examples, Sasha could build the infrastructure to support the idea in a physical space, but Sasha does not have the knowledge and does not want to buy or configure all the required infrastructure. Sasha's new idea is good but requires implementing the service and making it available on a virtual server (as in previous sections). The implementation entails hiring trained and experienced personnel to develop and test the system. In addition, this staff could build or hire accessory services to integrate to implement Sasha's idea. An example of this integration is Mailchimp [22], an online service that can manage follow-ups of newsletters and mailing campaigns sent to stakeholders through a personal or corporate brand email address. Among its features, Mailchimp enables the importation of contacts with whom you want to communicate by email, segmentation of contact lists and insertion of forms into emails. Mailchimp can also be integrated with more than 200 services in different business categories (Table 2.8).

As illustrated by the Mailchimp example, the costs vary depending on the requirements for the use of a service. The use of this email delivery service, like any other service that *MyBillProcess* uses, establishes a cost (usually monthly) that

Table 2.8 Comparison of costs for several Mailchimp options

Prices of Mailchimp [22]		
	Cost	Users
Free	$0/month	1–2000
Essentials	$9.99–$259/month	500–50,000
Standard	$14.99–$499/month	500–100,000
Premium	$299–$1099/month	10,000–200,000

Sasha will have to include in the cost of her service (usually monthly as well). The success of Sasha's idea will be assessed in terms of innovation, the savings generated by the increased productivity by avoiding bureaucratic burden in the lab, and the funding obtained through *MyBillProcess*.

2.2.5 Sasha Needs Mobility

Sasha is never at home but travels all year while working. Sasha cannot store all the information required for work on a laptop or mobile device and needs to be able to access relevant business information such as contacts, projects, contracts, and data everyday and continuously. Sasha also needs to be able to communicate with partners and customers. Sasha has hired an international Internet provider that has roaming agreements with the countries where Sasha works. By means of the service provider, Sasha is connected to the Internet at all times and can access all the necessary information to work. The cloud provides constant access to the required information and enables information exchange with collaborators and customers. The Sasha in any of the previous scenarios could also never be in the office; therefore, choosing cloud computing to manage daily work is the most reasonable option.

2.3 Everyone Is Sasha

In all the previous scenarios, a large portion of the solutions proposed in each case are interchangeable. The only characteristic that changes is the size of the cloud service required. All the factors involved in the decision to use the cloud can be summarized into two considerations: practicality and price. The practicality and suitability of cloud computing is beyond doubt for many kinds of daily work, and in some cases, it could be considered a waste to not opt for the cloud instead of maintaining comparable in-house systems. The use of cloud computing greatly reduces costs in many cases, especially for small and temporary uses, compared to the maintenance of in-house infrastructure that entails continuous and fixed costs. In an organization where revenues are not continuous and regularly guaranteed, installation and fixed maintenance costs are difficult to support.

Of course, Sasha always has both options. It is possible to build the hardware and software infrastructure necessary for the business, which entails a large initial economic outlay in most cases. Not only must the cost of hardware be considered but also the cost of the software and the personnel necessary to install it, configure it and keep it operational. The other option is to hire the same services in the cloud. In this way, the initial costs will not be as high and the investment may change depending on the needs of the evolving business.

Finally, cloud computing enables Sasha to establish economic control of the services that are contracted depending on the demand for those services and to work in a practical way to achieve the desired objectives.

References

1. Google Drive Cloud Storage (2019) https://www.google.com/drive/. Accessed 30 Oct 2019
2. IDrive (2019) https://www.idrive.com/pricing. Accessed 30 Oct 2019
3. Mega Cloud Storage (2019) https://mega.nz/. Accessed 30 Oct 2019
4. One Drive Cloud (2019) https://onedrive.live.com/about/en-gb/. Accessed 30 Oct 2019
5. Box Cloud Storage (2019) https://www.box.com/en-gb/pricing/individual/. Accessed 30 Oct 2019
6. ICloud Cloud Storage (2019) https://support.apple.com/es-es/HT201238/. Accessed 30 Oct 2019
7. GitHub (2019) https://github.com/pricing#feature-comparison/. Accessed 30 Oct 2019
8. GitLab (2019) https://about.gitlab.com/pricing/. Accessed 30 Oct 2019
9. Bitbucket (2019) https://bitbucket.org/product/pricing/. Accessed 30 Oct 2019
10. Skype (2019) https://www.skype.com/. Accessed 30 Oct 2019
11. GoToMeeting (2019) https://www.gotomeeting.com/es-es/meeting/pricing-ma/. Accessed 30 Oct 2019
12. Zoom (2019) https://zoom.us/pricing/. Accessed 30 Oct 2019
13. Webex (2019) https://www.webex.com/es/pricing/index.html. Accessed 30 Oct 2019
14. BACKBLAZE (2019) https://www.backblaze.com/backup-pricing.html. Accessed 30 Oct 2019
15. CARBONITE (2019) https://www.carbonite.com/backup-software/buy-carbonite-safe/. Accessed Oct 2019
16. CRASHPLAN (2019) https://www.crashplan.com/en-us/pricing/. Accessed 30 Oct 2019
17. Livedrive (2019) https://www2.livedrive.com/ForHome/. Accessed 30 Oct 2019
18. IDrive (2019) https://www.idrive.com/. Accessed 30 Oct 2019
19. WebsiteBuilderExpert (2019) https://www.websitebuilderexpert.com/building-websites/how-much-should-a-website-cost/. Accessed 30 Oct 2019
20. OuterBox (2019) https://www.outerboxdesign.com/web-design-articles/ecommerce_website_pricing. Accessed 30 Oct 2019
21. Hosting Facts (2019) https://hostingfacts.com/best-vps-hosting-review/. Accessed 30 Oct 2019
22. mailchimp (2019) https://mailchimp.com/. Accessed 30 Oct 2019

Chapter 3
From the Beginning to the Future

3.1 What Is Cloud Computing?

As noted in the past, the paradigm of cloud computing and all the ideas, variations and acronyms that it involves, constitute a single approach, a change in the location of computing resources [1]. Computing resources are no longer on your desk, your office or computing room, they are in a completely different location, possibly on the opposite side of the world and maybe onboard satellites in the future.

This transition has occurred during the past 10 years because of changes in the manner of understanding computing, initially focused on providing elastic and responsive services for websites with high traffic, and because of the need to have desktops synchronized so that a person can work anywhere and everywhere. These needs have evolved to many other realms and are now becoming mainstream. These characteristics shape the different options and approaches to cloud computing, which we will now discuss.

3.2 Types of Cloud

First, although Sasha has a very specific profile and the solution that provides the best fit has been explored as a case study in the previous chapters, no cloud computing solution is necessarily better than another. Multiple comparisons can be performed; however, the combination of provider and cloud model that best fits a specific purpose is unique. Chapter 4 will discuss the different considerations to take into account when selecting a cloud computing provider. Many variants in the realm of cloud and serverless computing have emerged as a result of how the needs of users and clients have grown. Therefore, a selected cloud solution will be a combination of solutions. These solutions are described next and are usually combined based on whether the application is commercial in character, the need for individual or shared use, and the security or level of infrastructure to be used (only hardware, hardware and software, etc.)

3.2.1 Commercial and Non-commercial

The co-existence of commercial and non-commercial options in the world of computers has existed since computers transitioned outside of laboratories to offices of corporations and homes. By the 1970s–1980s, such separation was also linked to a new paradigm, proprietary vs. free software. This distinction continues today, and cloud computing reflects a similar environment, where commercial or non-commercial cloud services can be obtained based on proprietary or free software. For those not familiar with the cloud environment, this scenario can be confusing, as sometimes jargon or names of technologies are used in a ubiquitous manner. In the next subsections, we explain the various types of cloud services and provide examples to contribute to a better understanding about what each one stands for, offers, and does.

3.2.1.1 Commercial Cloud

Almost every large company in the field of information technology is currently selling services related to cloud or serverless computing. Next, we discuss some of the basic options on the market and use them as an example of existing commercial

3.2 Types of Cloud

providers. Some of these solutions or companies are better than their competitors in terms of price or suitability, depending on the business of the client.

- **AWS:** the cloud computing solution provided by Amazon. AWS was launched in 2006 and can be considered to be the first cloud computing platform, at least in the way that we currently understand the cloud. AWS is usually considered the leader in terms of market share and is especially popular with those demanding high elasticity, that is, the capacity to use and adjust a large quantity of resources according to peak demand periods. A good example of typical users are companies or services that could experience unexpected peaks in demand or internet traffic because their contents suddenly go viral.

    AWS is also popular because of its so-called spot instances, that is, the possibility to use cloud computing resources almost for free when a given CPU is not busy. In such cases, AWS offers its use at a price near zero, and a given user can choose to run a task only if the price is below such a threshold. In this way, both parts receive a benefit: the user can perform a computational task at a very low price and AWS has a lower percentage of idle CPUs.
- **GCP:** the cloud platform provided by Google. Among the big names, GCP was the second platform to be launched in 2008. GCP is popular for scientific uses, such as the Large Hadron Collider (mentioned later in this chapter). To date, Google has not been entirely successful in making its cloud platform popular, and GCP is usually regarded as third in market share.
- **Azure:** the cloud computing service offered by Microsoft. Azure was launched in 2010 and has experienced considerable growth in recent years. Microsoft has acted as a market chaser to get positioned in the cloud and serverless computing business and appears to be doing well.

    Azure is oriented primarily to users that are already using other Microsoft products and simply want to deploy them in a cloud environment.
- **Rackspace:** born as a simple web hosting provider at the end of the 1990s and has evolved as a cloud computing company with a decent market share. Rackspace has contributed to some well-known technologies in the field of cloud computing and has been successful in capitalizing its experience in the sector.
- **IBM:** one of the main providers of technological services. IBM has been attempting to play a more important role in cloud computing in recent years. However, the market share and the development of its cloud business have not kept pace and reached a necessary level of competence. IBM finished the acquisition of the free software company Red Hat in the middle of 2019. Red Hat has a strong background of providing cloud services, and its integration should improve the position of IBM as one of the main players in cloud computing.

3.2.1.2 Non-commercial Cloud

As the popularity of cloud computing has grown in recent years, in some cases, users have viewed the cloud as a potential solution for daily work. However, some users prefer to not rely on a commercial provider, which has led to an alternative, the non-commercial cloud. Non-commercial cloud computing usually corresponds to the deployment of a cloud computing environment in a data center owned by the user. This scenario is related to a topic discussed in the next section, the "Private Cloud" model. Beyond this component of owned hardware, a non-commercial cloud also has a software component. Indeed, non-commercial cloud software can be the basis of a commercial cloud computing environment. Several options exist, and sometimes the differences between them, in terms of both philosophy and technical capabilities, are minimal. To identify the most convenient choice to deploy and to determine whether to use a cloud computing or serverless environment requires a high level of technical insight and is outside of the scope of this book. However, knowledge of the major names and basic ideas can be useful to understand what is available in the field.

One of the names to highlight is **OpenNebula**, which was initiated as a project in 2005 and first released in 2008. OpenNebula has since been very popular as a community-driven effort available under a free software license. However, although OpenNebula continues to grow in several aspects, its use is minor if we consider the whole market. Another major name is **OpenStack**, launched initially in 2010 by Rackspace and NASA, which differs from OpenNebula with respect to several technical characteristics. However, the most striking is philosophical. OpenStack is a free software solution used to deploy cloud computing environments, but it is not developed by a community of users but by a consortium of commercial providers or vendors that use it as a solution for the services they offer. Most of the vendors use or offer OpenStack among their solutions for cloud and serverless computing. Governmental research centers commonly use OpenNebula or OpenStack as the solution to deploy in-house cloud computing infrastructure.

Another long-standing idea is the possibility of mixing cloud computing with volunteer computing, which can be achieved in several ways. The most straightforward approach is to build a cloud computing service using computers from volunteers at their homes as the infrastructure [2]. One example is **Cloud@Home**, which is related to what is called "citizen science".

3.2.2 Cloud Models

Several ways of using and providing cloud computing services exist, most often related to security issues or privacy. The most commonly used cloud computing model is the **Public Cloud**, a model in which the servers in a data center can be shared by several users according to needs.[1] This approach is a flexible way to obtain access to resources, and the infrastructure is typically managed by the IT staff of the data center owner.

---

[1] This is independent of whether the data center is owned by a vendor or a private or public body.

In other cases, a given client or user might require additional security or privacy, making it undesirable to share the same servers or data center with other users. Such cases, where a single user occupies a full cloud computing service, represent a **Private Cloud**, a model where the infrastructure can be managed directly by the IT department of the client or by the vendor, according to needs. Moreover, a private cloud can be implemented by sharing a data center among several clients but with additional levels of security for each client, including firewalls, exclusive use of some servers, etc.

Experience shows that the best solution is sometimes a mix of both models; that is, Sasha could find a public cloud to be suitable for some tasks but require a private cloud for others. Therefore, the ideal solution is a **Hybrid Cloud**, a mix of a public and a private cloud.

These three models work around a single but important concept: property. Clearly, cloud computing can be used simply by a tenant, that is, using a data center provided by a vendor, in which case the user is not concerned with purchasing and maintaining hardware. The other option is to continue managing and maintaining your own data center. Each choice has its advantages and disadvantages for each user scenario. Notably, private cloud environments are typically sufficient for most users, including governments. An example is one of the biggest contracts in the defense sector granted in October 2019 by the US Department of Defense (DOD) to Azure that is worth up to 10 billion dollars.[2]

3.2.3 Types of Service

Cloud providers can offer different services from the perspective of the service abstraction, as shown on the diagram of hierarchy presented in Fig. 3.1. Higher abstractions (going up on the diagram) mean less operational load and more out-of-the-box features but also more constraints and more vendor dependency (lock-in). Given that every new layer goes on top and it is built upon the previous layer, abstractions, technologies services, and solutions based on the highest level of abstractions are usually newer. Going into detail across the layers of the diagram we find the following:

---

[2] At the moment of writing this book, AWS has reported that it will challenge the decision of the DOD of granting the contract to Azure.

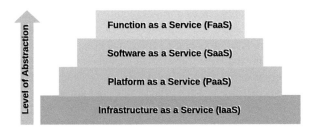

Fig. 3.1 Diagram of the hierarchy of cloud services

- **Infrastructure as a Service (IaaS):** The lowest layer of the hierarchy closest to the hardware. At this level, providers offer access to virtualized hardware. An example is AWS EC2 instances, which offer full access to the hardware of a virtualized instance.
- **Platform as a Service (PaaS):** The next level in the hierarchy offers another step toward abstraction and, usually, provides a full platform for users while hiding low-level details, such as the hardware. An example is Google Dataproc, which offers fully managed Hadoop and Spark clusters, where the users do not need to manage or configure the underlying infrastructure (including the network, instances and storage), and is presented as a single platform with very well-defined workflows.
- **Software as a Service (SaaS):** The line between PaaS and Saas is blurry and varies depending on the author (level of abstraction), but we can clearly state that services such as Dropbox or Gmail are SaaS. Such services and applications are fully hosted and managed by a third party (that does not necessarily need to be a cloud provider) and allow users to modify only predefined parameters (disk space, bandwidth, etc.) but not to access the underlying hardware resources or even be aware of them. Consider Dropbox, where users can manage objects (folders and files) and see how much free space they have available, but the level of knowledge of the infrastructure ends there for the user (for example, Sasha does not know and will not have to worry about what storage technology is used).
- **Function as a Service (FaaS):** More often known as *Serverless* computing, this level of abstraction is more directed to people who want to do development. In this scenario, the hardware resources are highly abstracted, so users can focus on writing software instead of managing the infrastructure (which is taken care of by the provider). FaaS is especially interesting for micro- and nanoservices [3], which are relevant for mobile device applications and the Internet of Things (IoT). Examples of FaaS include AWS Lambda and Google Cloud Functions. *Serverless* computing is described in more detail in Chap. 7 of this book: *Serverless experiments in the Cloud*.

3.3 Features of the Cloud

The are many reasons to use cloud and serverless computing technologies in research. Some are very well known and are discussed by vendors on their webpages for marketing purposes. These reasons are briefly discussed in the following.

3.3.1 Elasticity and Scalability

The scalability of cloud computing is very useful for research purposes in two typical scenarios. The first is related to users. Sasha might produce a dataset that is too large or that must be shared many times, making it infeasible or undesirable to manage in their own data center. In such cases, making the dataset available in a cloud system, possibly mirrored in several data centers around the world, might be the best choice, as such a system could handle the task without concerns about the number of users or bandwidth required for data transfer.

The other typical case is when Sasha needs to perform a computation in a limited amount of time and the local HPC resources are not sufficient. For such cases, Sasha can use cloud computing services to achieve the elasticity necessary to escalate the task and complete the computation on time. Typical scenarios for this scenario include the need to perform, in a very short time span, an unforeseen task because part of your research has suddenly become relevant for media or policy issues or because your client, founder or stakeholder requires it. In such cases, Sasha can use HPCaaS in the cloud to complement their local supercomputer or to avoid the queue to gain access.

3.3.2 Self-service Provisioning

Cloud self-service gives a user the freedom to request any available resources (such as instances or storage) from the catalogue of the provider. The providers sell various transparent automations that provide abstraction. It must be noted that in the cloud, Sasha does not provision a server (an instance) but must select a predefined image of a server that meets or is close to the requirements (for example, an image with Apache and PHP for a web server). In a more (computational) traditional environment, the available resources are predefined and sometimes shared (for example, when there is a queue to access CPU time), and self-service is not an option or requires the intervention of an operator.

As the freedom of choice increases, users must have a better understanding of the options and applicable technologies; therefore, most providers have created specific trainings and certifications (some for free), which is a business by itself.

3.3.3 Application Programming Interfaces

An application programming interface (API) is a (very) high-level abstraction that provides a standard way to communicate with software or infrastructure. In a very naive way, we can say that an API is software that provides a standardized interface through which other software can make calls to perform a desired action.

We will go deeper and provide examples for APIs in Chap. 6 of this book, but for now, we will say that APIs are *first-class citizens* on the cloud and represent the best practice on how to communicate between components. For example, creating a new directory (internally an object) in cloud storage is performed by calling the correspondent API of the service (e.g., S3 on AWS) and providing required attributes, such as the name of the directory.

Another example is the command line tooling that providers make available to their users, which tend to be wrappers that communicate with APIs and abstract common actions, for example, creating a new instance.

3.3.4 Performance Monitoring and Measuring

The cloud provides *observability*, which means that Sasha can access metrics and logs that provide the ability to make empirical observations using historical data of the available systems and services (and also to design models for applications such as resource usage forecasting). Examples of these metrics are CPU usage and free disk space at the instance or cluster level and more important characteristics such as error rate on the number of requests that a web server receives.

Each provider has a unique way of presenting this information using their own solutions (such as AWS CloudWatch or GCP Stackdriver), but free software projects are also available (such as Graphite [4] and Grafana [5]) and can be adapted to run on most clouds.

3.4 Cloud Native Concepts

All the concepts described in this section are required (or facilitate) to understand some parts of this book, and we encourage the reader to research and learn more about them. An in-depth discussion of the implementation of these concepts is not within the scope of this book; therefore, we introduce these concepts from a high-level perspective.

3.4.1 Cloud Native

Cloud native is a broad concept. As the reader can probably understand at this point, the delimitations are somewhat blurry in the cloud computing world and usually spark vivid arguments. Being *cloud native* means that an application or service takes advantage of, and to some extent requires, the environment that cloud computing provides. Moreover, the service or application very likely will not make sense, or even run, outside the cloud environment as most of the features (APIs, runtimes, etc.) will not be available or will not behave the same way.

To promote this concept and facilitate the creation of an ecosystem, the Cloud Native Computing Foundation (CNCF, which is part of the Linux Foundation) (https://www.cncf.io/) was created in 2015. Here, we use their definition of *cloud native*, as follows:

> Cloud native technologies empower organizations to build and run scalable applications in modern, dynamic environments such as public, private, and hybrid clouds. Containers, service meshes, microservices, immutable infrastructure, and declarative APIs exemplify this approach.
>
> These techniques enable loosely coupled systems that are resilient, manageable, and observable. Combined with robust automation, they allow engineers to make high-impact changes frequently and predictably with minimal toil.
>
> The Cloud Native Computing Foundation seeks to drive adoption of this paradigm by fostering and sustaining an ecosystem of open source, vendor-neutral projects. We democratize state-of-the-art patterns to make these innovations accessible for everyone.

3.4.2 Containers

Formally, a container is a virtual runtime environment that runs on top of a single operating system (OS) kernel and emulates an OS rather than the underlying hardware [6]. The most commonly used and well-known (PaaS) software that runs containers (via OS-level virtualization) is Docker [7]. Containers are considered to be core artifacts that power the concept of cloud native promoted by the CNCF.

From the perspective of scientific research, containers have been proposed as a potential solution to a long-standing and challenging problem: the computational reproducibility of scientific results. Computational tools are widely used in every field of research, and the reproducibility of results has become an increasingly important problem [8–10]. Cloud computing adds a layer of complexity, but a good description and use of containers could be a great tool to solve this problem [11–13], as all the software used is clearly described and available.

3.4.3 Microservices

Microservices are defined as decoupled, small (usually representing a single piece of business functionality), and independent services. For other authors and in different contexts, microservices can have a radically different definition, in the same way as for other cloud computing concepts.

Microservices are the method of writing software recommended by the CNCF, and we will consider a simple microservice (serverless computing) example in Chap. 7 of this book.

3.4.4 Kubernetes

Kubernetes is a free software container orchestration system created by Google that provides a set of abstractions to deploy, scale and run code. The most common use of Kubernetes is as a PaaS (for example, over GCP, AWS, or Azure), and it is widely used by the technology industry.

Kubernetes is developed and maintained by the CNCF and is the recommended method for orchestrating microservices over containers to fulfill an immutable approach to the infrastructure.

3.5 Environmental Issues

One of the issues with running large supercomputers or data centers is the environmental footprint, that is, the power consumption to run the infrastructure, including cooling. This factor must also be considered for cloud computing and is of special interest to organizations that want to reduce the environmental impact. Logic says that a larger centralized data center should be easier and cheaper to maintain than multiple separated computing infrastructures. Indeed this is true. Moreover companies providing cloud computing services are aware of the environmental problems associated with their business and make great efforts to limit them, for example, by attempting to be carbon neutral. The environmental impact can be an appealing argument when considering a transition to cloud computing; however, this is not a universal truth. For example, if you intend to migrate the software that you usually run on your PC (laptop, workstation, etc.) to the cloud, the savings are clear, as you can use a computer that consumes less power. However, if you are using cloud computing to process large amounts of data that involves continuous transfers, the environmental footprint of these processes plus the regular consumption of the data center might be worse [14]. However, when measuring these side effects, several factors must be considered: avoidance of in-house premises (local data center), all the associated energy costs (sometimes measured using popular metrics such as power usage effectiveness (PUE)), etc.

Some available data can shed light on this problem. A study by Microsoft on the uses of Azure found that the power savings, compared to in-house computing, when using SaaS are on average larger (greater than 90% in some cases) than those related to IaaS (52–79%) [15]. Also more than half of the energy used by AWS in 2018 was of renewable origin according to the data provided by the company itself and its carbon footprint compared to that of the average business data center is 88% lower according to 2019 data.

Last, it should be noted that data centers are much more efficient in terms of power consumption than they were years ago, and this situation is expected to continue to improve. Energy efficiency in the dedicated data centers provided by some vendors is continuously monitored, whereas for some in-house data centers, energy efficiency is considered only when they are designed or renewed. Therefore, when considering the environmental benefits of cloud computing, both the specific situation at the time of adoption and the evolution over time must be taken into account.

3.6 Success Stories of Cloud and Serverless Computing in Science

In recent years, cloud and serverless computing has reached a high level of maturity with success stories of its application in scientific research. Case studies can be used to illustrate potential applications; here, we briefly describe some representative examples. Furthermore, the web pages of the main providers list success stories, including some of the ones presented here. Clearly, this list is not comprehensive, and no doubt many more examples will have appeared between the time we are writing this book and you are reading it.

3.6.1 Archeology

The Stanford Archeology Center uses cloud computing to create accurate records of excavated items in a coherent way and to share them with the research community. Additionally, the Center has recently proposed a cloud computing solution to improve the archeological research of grand sites that includes the ability to combine IaaS, PaaS, and SaaS solutions [16].

3.6.2 Computer Sciences

A project on machine learning and data analysis from the University of California Berkeley is attempting to improve the efficiency of power consumption for batteries. Cloud computing has also been used to improve the chip design process, increasing its performance by 15%. Another example is a research group at the National University of Taiwan that has used cloud services as an expansion to its local cluster.

3.6.3 Energy

Cloud computing is already being combined with data mining and AI technology to forecast energy prices and to achieve a better match between electricity demand and generation. Moreover, there are tools that combine climate and energy data retrieved from cloud services, such as the EU Copernicus, to provide useful statistics [17].

3.6.4 Geosciences

In this field, cloud computing has been applied to data visualization, data storage and HPCaaS (including weather forecast systems and climate modelling). Examples exist for the subfields of oceans, atmospheric sciences, and climate modelling [18–20]. The FMI, NOAA, NCAR, and Met Office use cloud computing to handle data in a scalable manner, for example, when they face peak connection demand, data retrieval during extreme weather events, and massive quantities of outputs from climate model simulations. Moreover, applications of BOINC on cloud services exist [21]. The NREL in the USA uses cloud computing in several projects to collaboratively build and share relevant energy databases.

The possibility of using neural nets to reproduce weather forecasts has been tested in serverless environments. Such cases require the adaptation of neural nets to run in a serverless environment (see Chap. 7).

3.6.5 Library Science

The preservation of documentation and its on-line availability to support other kinds of research are of utmost importance. This is especially relevant in the case of old documents, where the use of digitized copies increases accessibility in a way that would not be possible because the original documents in some cases are in a very poor state of conservation. This kind of resources comprises high-resolution images and associated metadata. Libraries, such as the Catalan National Library, use cloud services to provide responsive access to consultation of these sources of information.

3.6.6 *Mathematics*

A good example of the application of cloud computing in collaborative research (and teaching) is CoCalc [22] (formerly known as SageMathCloud), a free web-based system that enables collaborative mathematical computations. CoCalc was developed using cloud computing credit schemes for research, such as those mentioned in Chap. 4. Thus, CoCalc is a really good example. According to Wikipedia, in 2017, CoCalc had approximately 300,000 active users.

From a broader perspective, the development of mathematical applications in cloud environments has been promoted by public bodies, for example, the MSO4SC project funded by the EU included HPC.

3.6.7 *Medical Sciences*

Numerous examples exist of the adoption of cloud and serverless computing for research in medical sciences. The FDA is currently using cloud computing to improve the processing of the massive quantities of manual reports on drug effects that they receive each year. Additionally, the CDC uses this technology to obtain and share information on disease evolution. In a rarer case of HPCaaS, the Walter and Eliza Hall Institute of Medical Research uses cloud computing to analyze high-resolution microscopy data. Furthermore, the Michael J. Fox Foundation has developed a research program where patients transmit data continuously from wearable devices to the cloud. These data can be analyzed together with data from other patients, representing a huge advancement over annotations from monthly medical exams and therefore benefiting the study of symptoms and drug research. Another example is the Ottawa Hospital Research Institute, which uses cloud computing resources as part of a program aimed to better understand why people are reluctant to vaccinate. Another example is research by the University of Western Cape that provides a web platform to study resistance to HIV drugs. Parallel workflows are also important and are used by the Smithsonian Institution to annotate genomes. Stanford University is using cloud computing to improve workflows in neuroimaging research. Furthermore, this technology is helping to improve computer security and compliance with strict regulations for medical information.

Several examples of serverless computing also exist. Examples of the use of AWS Lambda include Station X's GenePool, which can be used to analyze genetic data, and Benchling, a company that develops and sells software for gene editing.

3.6.8 Physics

Probably one of the brightest examples of applications of cloud computing technologies in physics is to manage the massive quantity of data resulting from the experiments with the Large Hadron Collider at CERN, an application that continues to adapt new technologies. An example of previous works is the use of GCP to analyze data from the so-called Atlas experiment. Another example is OpenFoam, a common free software in the field of computational fluid dynamics that can be directly run on AWS.

3.6.9 Social Sciences

An example in this field is the Queensland University of Technology Digital Observatory, which uses cloud computing technology to collect social media data around Australia and store the data for easy access by researchers. This use is highly beneficial from the perspective of analytics and visualization.

3.6.10 Space Research

ESA, NASA's JPL, and the University of Alaska use cloud computing to process and make available massive amounts of satellite images. In the case of the JPL, these images have been used for guidance systems of robots and to provide scalable video streaming of the "Mars Curiosity" mission. Moreover, serverless computing has been used to process astronomic images [23].

References

1. Hayes B (2008) Cloud computing. Commun ACM 51(7):9–11. https://doi.org/10.1145/1364782.1364786
2. Distefano S, Puliafito A (2012) Cloud@Home: toward a volunteer cloud. IT Prof 14:27–31. https://doi.org/10.1109/MITP.2011.111
3. Zimmermann O (2017) Microservices tenets. Comput Sci Res Dev 32(3):301–310. https://doi.org/10.1007/s00450-016-0337-0
4. Graphite (2019) Graphite. https://graphiteapp.org. Accessed 6 Nov 2019

References

5. Grafana (2019) Grafana. https://grafana.com/. Accessed 6 Nov 2019
6. Firesmith D (2017) Virtualization via containers. https://insights.sei.cmu.edu/sei_blog/2017/09/virtualization-via-containers.html. Accessed 6 Nov 2019
7. Docker (2019) Docker. https://www.docker.com/. Accessed 6 Nov 2019
8. Añel JA (2011) The importance of reviewing the code. Commun ACM 54(5):40–41. https://doi.org/10.1145/1941487.1941502
9. Añel JA (2017) Comment on "Most computational hydrology is not reproducible, so is it really science?" by Christopher Hutton et al. Water Resour Res 53(3):2572–2574. https://doi.org/10.1002/2016WR020190
10. Stodden V, Seiler J, Ma Z (2018) An empirical analysis of journal policy effectiveness for computational reproducibility. Proc Natl Acad Sci 115(11):2584–2589. https://doi.org/10.1073/pnas.1708290115
11. Boettiger C (2015) An introduction to Docker for reproducible research. SIGOPS Oper Syst Rev 49(1):71–79. https://doi.org/10.1145/2723872.2723882
12. Kurtzen GM, Sochat V, Bauer MW (2017) Singularity: scientific containers for mobility of compute. PLoS One 12(5):e0177459. https://doi.org/10.1371/journal.pone.0177459
13. Kim YM, Poline JB, Dumas G (2018) Experimenting with reproducibility: a case study of robustness in bioinformatics. GigaScience 7(7). https://doi.org/10.1093/gigascience/giy077
14. Baliga J, Ayre RWA, Hinton K, Tucker RS (2011) Green cloud computing: balancing energy in processing, storage, and transport. Proc IEEE 99(1):149–167. https://doi.org/10.1109/JPROC.2010.2060451
15. Microsoft (2018) The carbon benefits of cloud computing: a study on the Microsoft Cloud, 25 p. https://www.microsoft.com/en-us/download/details.aspx?id=56950
16. Wu Y, Lin S, Peng F, Li Q (2019) Methods and application of archeological cloud platform for grand sites based on spatio-temporal big data. ISPRS Int J Geo-Inf 8(9):377. https://doi.org/10.3390/ijgi8090377
17. Goodess CM, Troccoli A, Acton C, Añel JA, Bett PE, Brayshaw DJ, De Felice M, Dorling SE, Dubus L, Penny L, Percy B, Ranchin T, Thomas C, Trolliet M, Wald L (2019) Advancing climate services for the European renewable energy sector through capacity building and user engagement. Clim Serv 16:100139
18. Vance TF, Merati N, Yang C, Yuan M (2016) Cloud computing in ocean and atmospheric sciences. Academic, San Diego. https://doi.org/10.1016/C2014-0-04015-4
19. Zhuang J, Jacob DJ, Gaya JF, Yantosca RM, Lundgren EW, Sulprizio MP, Eastham SD (2019) Enabling immediate access to earth science models through cloud computing: application to the GEOS-Chem Model. Bull Am Meteorol Soc 100:1943–1960. https://doi.org/10.1175/BAMS-D-18-0243.1
20. Añel JA et al (submitted) Evaluation and intercomparison of cloud computing solutions for climate modelling
21. Montes D, Añel JA, Pena TF, Uhe P, Wallom, DCH (2017) Enabling BOINC in infrastructure as a service cloud systems. Geosci Model Dev 10:811–826. https://doi.org/10.5194/gmd-10-811-2017
22. CoCalc (2019) CoCalc. https://cocalc.com/. Accessed 6 Nov 2019
23. Malawski M, Gajek A, Zima A, Balis B, Figiela K (2019) Serverless execution of scientific workflows: experiments with HyperFlow, AWS Lambda and Google Cloud Functions. Future Gener Comput Syst. https://doi.org/10.1016/j.future.2017.10.029

Chapter 4
Show Me the Money

4.1 Money for Research Using Cloud Computing

In recent years, with the aim of promoting the use of cloud computing, several vendors have established programs to grant money or credit to researchers intending to test the viability of this option for their work. This approach has been, and continues to be, a win-win approach, as the scientific community benefits by gaining access to massive amounts of resources for HPC or data mining and vendors learn about the problems and needs of future potential clients, addressing them in advance and therefore developing a more robust and complete product. This approach also acts as a great marketing campaign, as the vendors are investing in projects for social benefit.

Some of these programs are or have been executed for a limited amount of time; others have been in progress for a longer time or are continuously open. Some examples are "Azure for Research" and "AWS Research Credits," and in some cases, there are calls for a specific topic, such as the Microsoft Climate Data Initiative or the Amazon Sustainability Data Initiative. In other cases, it is possible to contact vendors to ask about the possibility of obtaining a grant to use their resources; some vendors are open to collaboration. No doubt other options deserve to be explored, and most times, answers are only a search away on the web pages of providers.

Cloud computing can also serve as a model to optimize and account for research funding. In all science disciplines, it can be difficult to gain access to the infrastructure and instrumentation necessary to perform research. When applying for funding to perform various tasks, researchers must often request money to purchase the required infrastructure and instrumentation, and in some cases, funding bodies can be reluctant to provide the needed funds because the purchased material lasts longer than the research project itself. This problem is usually solved by obtaining funding from a different source, for example, core funding from research centers or shared-use critical infrastructure. However, in the case of computing infrastructure for punctual use (a single research project), as stated previously, access and use can be funded using ad hoc money included in a research grant, without having to maintain infrastructure purchased for a single purpose or having to justify spending money on something that lasts longer than the specific research being conducted.

4.2 Sharing Costs

We have previously discussed the case in which Sasha's laboratory is involved in research that requires massive HPCaaS and produces massive amounts of data that must be stored and how the laboratory probably must fight to obtain the required storage space and a proper Internet connection to share their data and results with their colleagues. This problem could arise if the research results obtained by Sasha become extremely popular or if the outcome of their research is a dataset that is widely used by the research community. Moreover, in such cases, cloud computing services could be a great way to manage the distribution of the results. If Sasha's colleagues simply want to download their data or produce results running their software, a cloud solution would simply require the laboratory to provide other researcher with authorization to perform these tasks via the cloud computing service. As an additional advantage, all the associated data transfer or HPCaaS expenses would be paid by the collaborators or stakeholders, not by Sasha's laboratory.

This characteristic represents a huge advance with respect to traditional schemes, where Sasha's laboratory might require their colleagues to sign a contract and pay for the service, to apply for a grant jointly or even be compensated with co-authorship for the outcomes of joint research. In summary, Sasha's lab would bear the expenses of producing the science but not the expense of sharing it: at a minimum, they can split the costs.

4.3 Selling Research

Cloud and serverless computing represent a major opportunity for businesses, a case that is partially developed in the previous section and results from the possibility of using cloud resources to provide access to a given research output. Consider Sasha after successfully developing a framework or software product for application and use. Many colleagues are likely to want to use Sasha's tool. One way to do so is by visiting Sasha's laboratory and collaborating. In some cases, such a tool might need to be adapted to the needs of the new user. One possibility in this case is to deploy the tool in the cloud and sell users access to the tool and customization services according to their needs. For example, this could be achieved through a spin-off company.

Another case is where Sasha's laboratory produces an amount of data that is too massive to process by their team alone and with potential uses that the team are not going to take advantage of. In this case, a good way to monetize the previous research work would be to make the data available in a cloud computing environment and let others download the data or perform cloud data mining by running algorithms over the dataset and charging a fee. This approach is followed by some of the private companies performing the R&D work presented in Sect. 3.6.

For Sasha, these approaches are a great way to enhance the use and impact of research and simultaneously reduce the burden that other research teams face when obtaining results using a system, methodology, or data from Sasha's laboratory. Moreover, this application presents some advantages for users with respect to the possibility to control the money spent on their research through a pay-as-you-go system.

4.4 What Happens If I Do Not Pay?

Sasha might fear the possibility of losing access to their research data if they do not pay the bills for the cloud computing services. First, this scenario applies only to commercial cloud providers that bill for the use of their resources. Beyond this situation, such fear is in some ways unjustified. However, simply maintaining data on the cloud has become almost free; the costs are mostly associated with data transfer and could be considered to be acceptable. Moreover, if Sasha needs to obtain access to the data but cannot afford it, Sasha could likely contact the provider and ask for the fee to be waived. Since the data are being requested for research purposes, it is highly probable that Sasha would be able to gain access, as we have already mentioned that specific grant programs exist for such cases, especially if there is no for-profit goal in the research work.

4.5 What Happens If My Provider Does Not Deliver

A different issue occurs when a commercial cloud computing provider fails to deliver and the user does not receive the service they have paid for. As with any other service, commercial providers are obligated to provide guarantees of service and are liable for failure to provide service. Commercial providers usually maintain high levels of transparency about technical problems and difficulties and grant credits to pay for unexpected issues. Moreover, in some cases, problems are solved rapidly and may not even be noticed by the user.

However, this is not the only potential issue. Indeed, the levels of reliability of cloud computing providers are extremely high. For example, during its main annual event on cloud computing in 2019 (Google Cloud Next), Google claimed to have the highest reliability among cloud computing vendors, showing that its service had been down for only approximately 200 min in 2018. The closest competitor was AWS, with a downtime of approximately 300 min. There was some discussion about how accurate these numbers are and who the most reliable vendor is, but the main point is that in an entire year, the cumulative downtime of commercial cloud computing vendors is in the range of 3–5 h. Clearly, this minimal downtime can still be an issue for critical systems, but the chances of a longer downtime in a full year when operating your own system are probably greater. For example, in the next section, we demonstrate how to assess a migration to cloud services based on real experience. The reliability of service is not considered, as in our view, this is not a necessary criterion for most users of cloud services.

4.6 Assessing Migration

In addition to simple tests and proofs of concept, you might need to evaluate if moving to a cloud computing infrastructure makes sense. In the following subsections, we discuss some considerations and aspects of how to assess a potential migration.

4.6 Assessing Migration

4.6.1 Selecting a Cloud Computing Provider

Selecting the cloud provider (or providers) is probably the most important decision [1], as it will not only define the possibilities (and limitations) but will also play a major role in next steps and in creating new opportunities for future work. Therefore, this step should be conducted carefully by fully considering all options.

This section does not propose new methodology for provider selection or even define which approach to choose; indeed, most vendors have their own documents on how to evaluate and perform migration to a cloud computing environment. We simply provide an understanding of the basic concepts so that the user can make the best selection for the problem they are trying to solve.

Although the assessment presented here is entirely agnostic, each vendor has a unique strategy and a target or niche of customers. Thus, the solutions designed and provided by each vendor are usually tailored to a given set of purposes. Therefore, the cloud provider that best fits your needs is likely to be one that is already used by researchers in your field.

4.6.2 Problem Definition

The first step is to identify the problem to be solved and its high-level components. The more specific we can be in defining the requirements, the easier it will be to find the best solution (and provider). For instance, in the example considered in this section, Sasha is attempting to run a climate simulation to calculate data such as cloudiness (how cloudy a day is) on certain parts of the Earth.

4.6.3 Gather Requirements

Different methodologies and approaches [2, 3] can be used for this selection, but the next step is to gather all the requirements to understand the use case and attempt to satisfy as many of the following dimensions as possible [4]:

- **Suitability**: The available services and systems are sufficient to cover the needs of our problem. These requirements include the system performance and the speed of the network connection between data centers of the provider to obtain mirrored copies of the data.
- **Usability**: Ease-of-use interfaces, APIs, etc. are adapted to the level of knowledge of the staff that will be managing the project.
- **Elasticity**: The provider is sufficiently flexible to provide different (or even better) options for our resource needs.
- **Costs**: The costs of the services from the provider are aligned with our expectations: whether waivers or discounts are available should also be considered.
- **Security and Compliance**: There is always certain level of security needed, which requires effort and planning. Sasha is aware that a lack of security can lead to major losses, and not only economic. The same applies to compliance, which includes adapting to regulations, running code in a more benevolent jurisdiction if needed, and ensuring that services comply with the law. If not planned correctly, these two dimensions can have serious consequences or directly lead to failure.

Sasha should assign a value from 0 to 1 to each dimension in terms of its importance, with 0 being the lowest value and 1 the highest.

Following the example from the previous section, after discussing the situation with the team, Sasha obtains a list of detailed requirements:

- The simulation has been running on a single on-site server, but the researchers would like to be able to perform multiple runs in parallel in the future. The full specifications of the server are available.
- The application is CPU demanding but does not need much memory.
- Input data are from a public API available over the Internet.
- Resulting data are small and should be made available to the team members via shared storage.
- The results are sensitive, so a specific level of security is required. Sasha's laboratory operates under compliance rules (and audits) from the government.
- The project is in the early stages so the budget is limited, but Sasha could receive more funding if the results are satisfactory.

Given these requirements and further discussion, Sasha's team has set the weights for each of the aspects to consider as shown in Table 4.1.

Table 4.1 Example of weights assigned to each feature to assess a migration to cloud computing

Suitability	Usability	Elasticity	Costs	Security and compliance
0.8	0.3	0.8	0.5	1

4.6 Assessing Migration

Table 4.2 Comparison of providers to assess a migration to cloud computing based on the weights assigned to each feature

Dimension	Weight	Provider 1	Provider 2	Provider 3
Suitability	0.8	7	5	6
Usability	0.3	9	8	3
Elasticity	0.8	5	5	8
Costs	0.5	4	7	6
Security and compliance	1	7	8	2
Total		32	33	25
Weighted total		21.3	21.9	17.1

4.6.4 Obtain the Data

After all the problem boundaries are defined, the data for the providers are assessed in our preferred format, for example, a table. Given our requirements, a weight is assigned to each dimension, and the providers are rated according to their ability to satisfy the requirements. For example, for the problem defined in the previous sections, we assign a value from 1 to 10 to the dimensions and adjust the value based on the previously defined weights. Table 4.2 shows an example of how to assess provider that best fits according to what Sasha needs.

4.6.5 Review and Decide

Once the data are obtained (as shown in Table 4.2), an initial evaluation is performed, and options that are obviously unsatisfactory (i.e., many of the requirements are not satisfied) can be discarded.

For our example, we must decide between Provider 1 and Provider 2. A new iteration can be performed by re-adjusting the weights based on the findings. If more than one provider ensures similar levels of service, we might have to look for more information about each service or feature to obtain more accurate scores. The following additional steps might be helpful to Sasha:

- Communicate with the sales channel for all the evaluated providers, as the standard pricing usually does not reflect various discounts or conditions available.
- Try to obtain demo time or credits to empirically assess whether the provider can meet the requirements.
- Discuss again the problem to be solved and see if new information and findings have altered some of the requirements.
- Iterate the process many times until the number of options is reduced (e.g., to only two plausible providers).

4.7 Conclusions

To conclude, always exists a variable in cloud computing: usage of a cloud computing service has an associated cost, and depending on your use case and field of work (the costs for educational purposes and business purposes may differ) you may have access to waivers. Overall, a well-designed cloud solution could save (or make) you money, but as detailed in this chapter, many aspects other than money must be considered.

Furthermore, when selecting a cloud computing provider, it must be taken into account that organizations increasingly use both "multicloud" and "hybrid cloud" approaches when working with cloud computing services. Multicloud means that an organization does not rely on a single vendor or provider for any of several different reasons, like taking advantage the different approaches that each vendor has to cloud computing (software, design, etc.). Another reason is to avoid the typical vendor lock-in. In the hybrid cloud approach, an organization works with both private and public cloud solutions but not necessarily different providers. Therefore, this solution might be considered if fidelity to a given vendor is the best option.

References

1. Misra SC, Mondal A (2011) Identification of a company's suitability for the adoption of cloud computing and modelling its corresponding return on investment. Math Comput Model 53(3–4):504–521. https://doi.org/10.1016/j.mcm.2010.03.037
2. Bildosola I, Río-Belver R, Cilleruelo E, Garechana G (2015) Design and implementation of a cloud computing adoption decision tool: generating a cloud road. PLoS ONE 10(7):e0134563. https://doi.org/10.1371/journal.pone.0134563
3. Etsy (2018) Selecting a cloud provider (code as craft post). https://codeascraft.com/2018/01/04/selecting-a-cloud-provider/. Accessed 10 Jun 2019
4. Repschläger J, Wind S, Zarnekow R, Turowski K (2011) Developing a cloud provider selection model. Paper presented at the enterprise modelling and information systems architectures: proceedings of the 4th international workshop on enterprise modelling and information systems, EMISA 2011, Hamburg, September 22–23, 2011

Chapter 5
Tools in the Cloud

5.1 Sasha Wants Better Tooling for Their Organization

As the leader of a research organization, Sasha wants to improve their tooling and service standardization while following the best practices and patterns for their work. One requirement is that the tools be made available over the network, so SaaS solutions are a good option (moreover, SaaS solutions do not require a dedicated IT team for support).

5.2 General Productivity

Without considering a specific area of application, a large variety of generalist software can be used for projects of various kinds (including scientific or technical work). In this section, we detail common categories and various examples.

5.2.1 Office

Office suites are among the most frequently used applications, with especially strong usage among scientists: paper writing (and sharing), spreadsheets for budgets, data, etc. The most common SaaS office suites are the following:

- **Google Docs/G Suite:** Google Docs is the office cloud suite created by Google that includes a word processor, a spreadsheet, and a presentation application. A free version is available to general customers, and a paid monthly subscription (G Suite) that provides extended features and storage is available for professionals and companies (https://www.google.com/docs/about/). G Suite has many users around the world. The multinational pharmaceutical company Roche is a good example of a private organization conducting research.
- **Office 365:** This is the cloud version of the prominent Microsoft suite that provides an online adaptation of Word, Excel, and PowerPoint. This solution requires a paid yearly subscription (https://www.office.com/).
- **Zoho Office:** The Zoho Office suite is a web-based solution that contains (among other software) word processing, spreadsheet, presentation, database, and project management software. A free (limited to teams of 25 users) version is available, in addition to a monthly per user paid version with increased available space and features (https://www.zoho.com/docs/).

5.2.2 Communication

Chat and general purpose communication software have been around for a while, but in recent years, collaborative communication tools and platforms have been gaining popularity and considerable traction from the community. The most commonly used platforms are the following:

- **Slack:** Slack is a collaboration-oriented tool rather than a pure communication platform with many well-known features (some of them directly adapted from IRC), as well as integrations with extended platforms or products such as Jira and GitHub. Slack also supports audio and video. The free version is suitable for small teams, and there is also a paid per user per month version (https://slack.com/).
- **Google Hangouts:** The communication platform from Google that is integrated with all their tools and platforms (such as G Suite) includes audio and video communications (https://hangouts.google.com/).
- **Zoom:** A video conference and call platform that provides chat rooms and capability to share content (https://www.zoom.com/).
- **IRC:** The Internet relay chat, which has been around since 1988, is a well-known protocol for communications. Many free clients are available for various platforms. IRC remains a valid solution because many cloud providers provide access (sometimes for free) to their IRC servers.

5.3 Scientific and Engineering

Cloud-hosted software for science and engineering has been a game changer in industry with the development of new models of business, such as paying for only the required features or resources, thereby improving cost optimization. This approach also allows users to access platforms or systems that they could not previously access. Some of the related software is discussed below.

5.3.1 Document and Paper Edition and Sharing (LaTeX)

In science, publishing results has, for centuries, been the main way to share knowledge and boost advancement. Several collaborative document edition approaches using cloud computing services have improved this task, mainly in cases where a scientific work is developed by several people or multiple teams. Indeed, this book has been written using these tools. Some examples are as follows:

- **Overleaf:** Overleaf provides a LaTeX collaborative writing and publishing platform. A limited free version is available for small and medium projects (but cannot have private or protected content), in addition to a yearly subscription option (https://www.overleaf.com). Among its features, Overleaf enables direct submission of a paper to the journal of your choice.
- **LaTeX Base:** Affordable LaTeX collaborative edition system that satisfies the requirements of most solo authors or small teams (https://latexbase.com).

5.3.2 Project and Issue Management

One of the many tasks of a scientist or engineer is to track their work and share their results with their peers in real time. Project and issue management SaaS are powerful tools to perform this task. The most popular platforms are as follows:

- **Jira Cloud:** Jira (and Atlassian) are probably the most well-known project and issue management software. The platforms enable users to perform tasks from addressing very simple issues to highly complex project management (including Agile dashboards and integrations). Another system provided by Atlassian is also worth noting. Trello (https://trello.com) offers a lean system (based on cards and very Agile oriented) for managing small projects or works and has a very good free version that provides complete integration with other Atlassian products such as Bitbucket. A monthly license (per user) is required (https://www.atlassian.com/software/jira).

- **Basecamp:** Basecamp is a suite of integrated tools (issue management, communications, etc.) for project management that requires a monthly fee, but waivers (and free versions) are offered for educational purposes (https://basecamp.com/).
- **Asana:** This platform, used by major IT players, is focused on timelines and tracking. The free version is suitable for small teams, and a monthly per user fee is required for access to the advanced features (https://asana.com).

5.3.3 Version Control

Version control processes aim to keep track of different versions of one or more artifacts (software, documentation, etc.) and are commonly used in the software development process and for documentation. The most common platforms are the following:

- **Git:** Most commonly used by the open-source community, Git is distributed and was originally created by Linus Torvalds for development of the Linux kernel. Git is offered as SaaS by many providers, including the following:
  - *GitHub*: GitHub offers free public repositories and is the central hub for the free software community (https://github.com/).
  - *GitLab*: A growing alternative to GitHub that has been gaining popularity in recent years by offering private repositories and advanced features for testing and building. Additionally, GitLab was raised as an attractive alternative to GitHub after the latter was purchased by Microsoft in 2018, which has produced some problems from the perspective of scientific reproducibility [1] (https://gitlab.com/).
  - *Bitbucket*: The implementation from Atlassian that interacts well with the Jira ecosystem (https://bitbucket.org).
- **Mercurial:** Introduced at the same time as Git, Mercurial is a monolithic approach that is less complex than Git and is used by many major projects (like Mozilla and Nginx) (https://www.mercurial-scm.org/).
- **SVN:** Apache Subversion (SVN), one of the oldest version control systems (it has been around since 2000), offers a more centralized approach than Git (https://subversion.apache.org/).

Although it is not the purpose (or even within the scope) of this book to discuss Git, knowledge of its basic functioning and workflow is highly recommended and is, generally speaking, an expected skill [2].

5.3.4 Other Relevant SaaS

As mentioned at the beginning of the chapter, it is impossible to cover all the products offered as SaaS. New applications come out every day, some of which are quite surprising. For example, AWS is working with universities to provide chatbots for students to ask questions related to the university using a smartphone, a tablet or Alexa device [3, 4]. An overview of other relevant SaaS for scientific and technical environments includes the following:

5.3.4.1 Integrated Development Environment

Some of the most relevant integrated development environments are as follow:

- *Codeanywhere*: A cross-platform IDE in the cloud that enables collaboration on the development of software projects using a web browser on any kind of device (https://codeanywhere.com/).
- *AWS Cloud9*: Similar to Codeanywhere but developed by AWS (https://aws.amazon.com/cloud9/).
- *Ideone*: Cloud tool to compile and debug source code in several programming languages (https://ideone.com/).

5.3.4.2 Diagram Design

Tools are also available to draw and make diagrams in a collaborative way in the cloud simply using a web browser. For the interested reader, we note three available tools: draw.io (https://www.draw.io/), yEd Live (https://www.yworks.com/yed-live/), and Lucidchart (https://www.lucidchart.com/).

5.4 Sasha Decision

After considering the existing tools and cloud environments and discussion with the team, given the budget and needs, Sasha must select a tool. A potential complete solution could be the following:

- **Document Edition and Sharing:** G Suite for general documents (and discussions) and Overleaf for LaTeX (and collaborative paper edition).
- **For Communication:** Slack because it appears to be the current standard, it is free and it has many useful plugins.
- **Task Management:** Trello because it is free and small; the team does not need a more complex product.

- **Version Control:** GitLab, which offers Git with private repositories (so the team can keep the code private until it is ready to release) and many free tools for testing and building.
- The use of other tools is still under discussion because the team must agree on which standards (and formats) to use and share.

References

1. Nature (2018) Microsoft's purchase of GitHub leaves some scientists uneasy. Nature 558:353 https://doi.org/10.1038/d41586-018-05426-0
2. Git gittutorial – A tutorial introduction to Git. https://git-scm.com/docs/gittutorial. Accessed 20 Jan 2019
3. Lancaster University (2019) Lancaster University launch pioneering chatbot companion for students. https://www.lancaster.ac.uk/news/lancaster-university-launch-pioneering-chatbot-companion-for-students. Accessed 10 Mar 2019
4. AWS (2019) Building a multi-channel Q&A chatbot at Saint Louis University using the open source QnABot. https://aws.amazon.com/es/blogs/publicsector/building-a-multi-channel-qa-chatbot-at-saint-louis-university-using-the-open-source-qnabot/. Accessed 17 Oct 2019

Chapter 6
Experiments in the Cloud

6.1 Sasha Is Ready to Start a New Experiment

Now that Sasha has a clear overview of what to expect from cloud technologies and, most importantly, how to obtain a budget and understand how to use it, Sasha is ready to begin a new experiment or project in the cloud. Since its genesis, one of the goals of the cloud has been to run software in a massively distributed manner, thereby providing ubiquitous and transparent access to any needed resource [1]. Running and writing software are, probably (along with interpreting results), the most common tasks of a scientist or engineer on a regular day. One of the main problems software developers face on the cloud is the series of frustrations resulting from not understanding (or even worse, trying to fight) the massively distributed nature of the Cloud, where the "out-of-the-box" resources are not always the same as those expected in a traditional HPC environment (e.g., low-latency network by default).

6.2 Running Code on the Cloud

Getting code to run in the cloud might appear, initially, to be trivial, but this is not necessarily true, especially if the software (and the infrastructure needed to run it) is highly complex. This same feature applies to scientific code.

6.2.1 Understanding Software Development on the Cloud

Before the cloud paradigm, most developers wrote and ran software via two classic approaches:

- **Using "local" resources:** ad hoc set of resources (e.g., workstations) that were easy and fast to access but highly static. If more hardware was needed, it had to be purchased, and sometimes available resources were not used at all (e.g., disk space or memory).
- **Using an HPC infrastructure (or provider):** developers or scientists from participating/affiliated organizations are granted access to a shared pool of resources (e.g., computing nodes) and can request access to higher (usually single tenancy) resources for specific tasks (e.g., running an experiment). Access to any resource is, usually, obtained via a resource manager/scheduler [2] and depends on the level of access granted. The available resources and scaling depend on the capacity of the provider.

Notably, these solutions are not free, and their costs are often abstracted or hidden from the users (direct funding or budget assignments) [3].

Cloud computing offers a radically different proposal in which resources are highly dynamic and you can use exactly what you need (thereby reducing idle resources). Clearly, there is no such thing as a "silver bullet," so despite all the conditions that we have already discussed in previous chapters, before jumping into cloud computing, Sasha should consider some questions before understanding how to run or write code for the cloud:

- **Is Sasha's software suitable for the cloud?** This question might appear to be trivial, but in reality, it is a hard blocker: your software (as it is) could not be easily run on the cloud. No one other than yourself (or whoever wrote the software) can really answer this question. The answer to this question is very likely to be "yes," but you should consider characteristics such as the hardcoded infrastructure and investment in resources in case parts of the code must be rewritten [4].

- **How many resources does Sasha *really* need?** This question is very important: Sasha should know beforehand which resources are needed (of course, a certain level of uncertainty is always acceptable). Basic questions to answer are the following:
  - How many CPUs and how much memory are needed (and how many per node)?
  - How much storage is needed? Does it need to be persistent (e.g., one bucket)?
  - How long is the application going to run?

  Addressing these questions in advance could save Sasha considerable trouble, such as hitting API limits or even worse, wasting funds on unnecessary resources. Moreover, Sasha should know the limits of their cloud provider, even though the cloud resources are usually more extensive than those of the current HPC infrastructure. This information could be obtained by contacting the provider.
- **What are acceptable thresholds for Sasha's software or experiments?** In other words, what are Sasha's expectations? Code or experiments can be run at a very low cost if there are no hard time constrains (e.g., you do not mind that it takes longer to run your code).
- **What is Sasha's budget?** Finally, all the previous points and questions are really aligned to answer this very question: How many resources are available? The more accurately the previous questions can be answered, the better the answer to this final question will be (please review Chap. 4 for further reference).

6.2.2 Application Programming Interfaces (APIs)

6.2.2.1 What Is an API?

APIs, very high-level abstractions that provide a standard way to communicate with software or infrastructure, are common in computer sciences. APIs provide a system access point or library function with a well-defined syntax that is accessible from application programs or user code to provide well-defined functionality. APIs play a major role in cloud systems and can have various forms (e.g., the POSIX API). The basic workflow of an API is shown in Fig. 6.1.

Modern cloud systems commonly use and provide representational state transfer (REST) via create, read, update, delete (CRUD) operations and/or GraphQL APIs [5] encapsulated on HTTP. Some interesting examples of this approach are as follows:

- One RESTful API is the NASA (Open) Earth API [6] that allows exploration of available datasets (including images).
- Several RESTful APIs from Google (e.g., search) are publicly available [7].

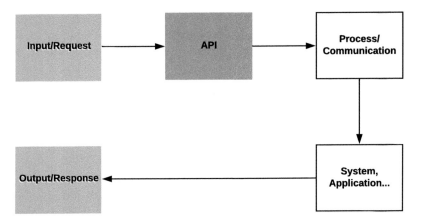

Fig. 6.1 Basic workflow of an API

6.2.3 Cloud Environment and Runtimes

The cloud itself as an entity, when talking about the degree of access or control, can be experienced as a spectrum: from raw access to hardware (via low-level APIs), to high-level APIs and complex SDKs. The spectrum is continuous: the Cloud is not a monolith but a *living ecosystem* composed of moving components that, usually, interact and communicate with each other. Therefore, even though the components are independent, they tend to create interdependencies via communication (usually via APIs). All the available components (services or systems), their exposed interfaces and variables constitute the cloud environment, which is the medium in which this ecosystem is immersed.

Understanding the Cloud environment is key to being successful in running or developing software for the cloud. However, there is no single cloud environment: every provider has its own environment, and it is recommended to research (and read the official documentation) available on every cloud environment.

6.2.3.1 Why Are APIs so Important in the Cloud?

As explained previously, being in the cloud means being immersed in its environment, which exposes its resources via APIs and a set of allowed operations. Other aspects to take into account are as follows:

- Many features are only available via API and must be implemented programmatically.
- APIs offer the opportunity to create repeatable or automatable patterns; the UI or any other methods are sub-optimal (and could be subject to change over time).

6.2 Running Code on the Cloud

The main public cloud providers have extensive well-defined documentation of their APIs (including examples for different programming languages):

- Azure REST API [8]
- AWS Guides and API [9]
- Google Cloud API [10]

6.2.3.2 Example: Using a Cloud API

As an example of how to use a Cloud API, we will consider how AWS EC2 makes the information of an instance available within itself via metadata in JSON format; that is, an instance can obtain information about itself by querying a local service that provides metadata. For example, the following command can be run to obtain a list of available items in the metadata:

```
$ curl http://169.254.169.254/latest/meta-data/
```

Given that list, the following command can be used to obtain the public IP(v4) address of the instance:

```
$ curl http://169.254.169.254/latest/meta-data/public-ipv4/193.146.46.94
```

Clearly, this process is one of the lowest levels of access. In this case, the interface can be accessed directly to make a GET request with curl. This approach is needed only in certain cases; if possible, it is recommended to use the SDK usually made available by cloud providers (or third parties). Higher levels of abstraction are possible when using the SDK (e.g., communicating with different APIs for different services), and writing code becomes easier. SDKs are available for the major public cloud providers:

- AWS SDK [11]
- GCP SDK [12]
- Azure SDK [13]

An added complexity is the concept of *Region* that exists for different cloud providers. A region is the geographical location where the infrastructure is physically located and the resources are run. The region is not only a physical concept but also indicate that the available resources, even their features, could be different. For instance, on GCP, the available CPU types in the region *asia-east1* are different than those available in *europe-north1*.[1]

When the issue of regions is added to the former example, the list of running instances for a project on GCP in Python requires the following steps:

---

[1] Note that names and regions change, so these names may not correspond to the Google Cloud Regions that are currently available when you are reading this book.

- Install the Google Cloud SDK as described on its web page (this step must be performed only once).
- Install the client library for Python (this step must be performed only once).

```
$ pip install --upgrade google-api-python-client
```

- Once the dependencies are met, this simple class will return the list of running instances:

```python
"""
    This code list the instances on a given project and zone.
"""
import os
from pathlib import Path
import googleapiclient.discovery

# Credentials can be generated at your:
# https://console.cloud.google.com/iam-admin/serviceaccounts

# Put your JSON credentials file on your $HOME/.google/credentials.json
os.environ[
    "GOOGLE_APPLICATION_CREDENTIALS"] = "%s/.google/credentials.json" % Path.home()


class GCPCompute(object):
    """
    This class contains the logic to return a list containing
    the instances for a zone (for a project).
    """

    def __init__(self, project: str, zone: str):
        self.compute = googleapiclient.discovery.build('compute', 'v1')
        self.project = project
        self.zone = zone

    def list_instances(self) -> list:
        """
        Return a list of items (instances).
        """
        result = self.compute.instances().list(
            project=self.project, zone=self.zone).execute()
        return result.get('items', [])


if __name__ == "__main__":
```

6.2 Running Code on the Cloud

```
PROJECT = "my-experiment"
ZONE = "europe-west1-c"

GCPCOMPUTE = GCPCompute(project=PROJECT, zone=ZONE)
INSTANCES = GCPCOMPUTE.list_instances()

print(">> Project : %s " % PROJECT)

# Just print instances names (if any):
if not INSTANCES:
    print("\nInstances (name) for zone %s:" % ZONE)
    for instance in INSTANCES:
        print("-> %s" % instance['name'])
else :
    print("\nNo instances found for zone %s" % ZONE)
```

- The output of running this code is as follows:

```
$ python list_instances.py

>> Project : my-experiment

Instances (name) for zone europe-west1-c
-> Instance_1
-> Instance_2
-> Instance_3
```

6.2.4 Code in the Cloud: Considerations

At this point, in terms of both running and writing software in the Cloud, Sasha must decide whether to start from scratch (and fully embrace the Cloud paradigm) or to refactor the code to run in the cloud, which implies the following:

- **Starting a cloud native project:** Architect the systems and implement them from scratch to take advantage of all the resources (such as APIs) while mitigating the pitfalls (e.g., poor performance in specific situations) of the cloud environment.
- **Making software and systems cloud enabled [4]:** Migrate as much of the software as possible to take advantage of the cloud or, in a more limited case, wrap the software so that it can at least run within the cloud (e.g., within a regular instance).

Next, Sasha must understand some other aspects before starting the implementation (for either of the two approaches) that will make Sasha aware of the limitations, trade-offs, and impacts of their work.

6.2.4.1 General Aspects

- **Most of the hardware is (highly) abstracted:** Sometimes the information the system provides does not reflect the reality (e.g., AWS EC2 Computer Units). This characteristic can make it difficult to fine-tune the results (e.g., specific compiler flags).
- **Some systems share resources:** The CPU or I/O (IOPS) can share resources, which can result in some unexpected results, such as CPU bursts or underperforming disks.
- **Always use APIs by default (and avoid using the UI):** Some resources are only available in this manner.
- **Take advantage of "out-of-the-box" systems and components:** technologies such as containers (e.g., Docker or Kubernetes) and PaaS (e.g., Hadoop as a Service) can save considerable time in setting up the infrastructure.

6.2.4.2 Network

- **Latency:** Network latency is one aspect that greatly impacts code, especially scientific code that takes advantage of technologies such as MPI and low-latency networks (e.g., InfiniBand). By default, most cloud providers offer a network intended for general purposes (even though some providers claim ultra-low latency) that could dramatically degrade the performance of scientific applications. Therefore, the general recommendation is to consider the available optimizations per provider (some offer HPC-optimized environments, separated from the regular deals) and conduct an initial assessment (e.g., perform an initial run to determine the impact on the application) or re-imagine the software in such a way that vertical scalability gains weight versus horizontal scalability and low-latency communications.
- **Limited troubleshooting:** In the cloud environment, troubleshooting issues with the network are limited given the nature of the networks and their abstraction.
- **Proximity and hardware allocation/affinity:** By default, the allocation of resources such as instances is limited, usually by region. This limitation can be challenging for applications that require low latency and high affinity and should be discussed with the provider, as they can offer on-demand and customized solutions to this problem (e.g., separated environments).
- **Private networks:** Some networks that need to be reachable from Sasha's systems or services might not be directly reachable by the cloud provider. For example, a private research facility might provide access to specific data (that

could be required for our experiments) over only their own private networks. A VPN is a potential solution but is not always available for all providers or locations.

6.2.4.3 Storage

- **Unlimited storage:** One of the main advantages of the cloud is the high level of abstraction from resources and their presentation, which makes storage virtually unlimited (limited only by the budget). This unlimited storage can be viewed as an advantage versus on-site infrastructure, where more storage requires a more or less complex set of operations (such as racking or configuring the new hardware).
- **Different levels of storage:** In addition to the previous point, abstraction can also provide access to different levels of storage, that is, from fast and expensive (for frequent access) to slow and inexpensive (for backups or infrequent access).

6.3 Starting an Experiment in the Cloud

At this stage Sasha, has sufficient knowledge and information to setup an experiment in the cloud. We reiterate various points for review:

- **Project management:** Although not mandatory, it is highly recommended, and good practice, to use a project management tool. Moreover, Sasha should follow a well-defined working methodology [14], which will increase the probability of success.
- **Version control:** Sasha should maintain and share code and documentation under a version control system (such as Git).
- **Select the cloud provider and technologies** to use (and try to calculate dependencies and costs).

6.3.1 Example: Month Classifying Experiment

Sasha's new project is simple software that classifies a given month in terms of cloudiness. Sasha plans to use available funds for research on AWS. In the initial naive approach, the source of the data will be NASA's Earth data API [6]. Specifically, the cloud provider and technologies that Sasha will need are as follows:

- **Cloud provider:** For this case, let us assume that it is AWS.
- **Technologies:**

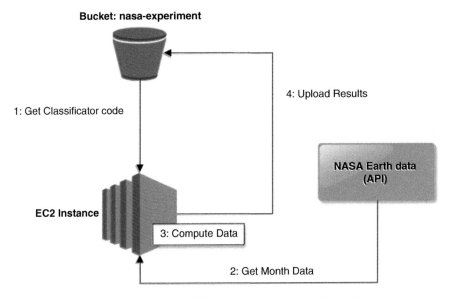

Fig. 6.2 High-level architecture and workflow for the cloudiness month classifier experiment

- Programming language and libraries: Python 3 and NumPy.
- Tooling: AWS Command Line Interface.
- Third party APIs: NASA's Earth data API.
- Cloud technologies/services: AWS EC2 (compute) and S3 (storage)

The software will be named *classificator.py*: a very high-level architecture and workflow is shown in Fig. 6.2.

- **Pre-requirements:**
    - This step assumes that a proper virtual private cloud (VPC) and a subnet are correctly created within AWS (more information on how to set up a VPC and subnet can be found in the official documentation [15]). For this example, the created subnet has the id *subnet-1a2b3c4e*.
    - A valid/created key pair is required to access the instance via SSH if needed (a key pair can be created on the AWS console (UI) EC2 Manager→Key Pairs→Create Key Pair). For this example, the key pair will be *nasa-experiment*
    - An S3 bucket is created (this can be done on the AWS console (UI) under S3→Create Bucket), and the classificator.py code is located in the bucket with public access (so it can be downloaded from the instance).

6.3 Starting an Experiment in the Cloud

This script will bring up a t2.micro instance on Ireland running Amazon Linux (ami-047bb4163c506cd98) on the subnet subnet-1a2b3c4e that is accessible via SSH using the key *nasa-experiment*.[2]

The first part of the code creates a basic infrastructure on AWS composed of a single instance from the offered free tier. This is a simple shell script that uses the AWS CLI.[3]

```
#!/bin/bash
# AMI id is for Amazon Linux
aws ec2 run-instances --region eu-west-1 --image-id ami-047
    bb4163c506cd98 --count 1 --instance-type t2.micro --key-name nasa
    -experiment --subnet-id subnet-1a2b3c4e --instance-initiated-
    shutdown-behavior terminate --user-data file://${PWD}/scripts/boot.sh
```

The content of the boot script "boot.sh" that installs the dependencies and runs the simulation code is as follows[4]:

```
#!/bin/bash
yum update -y
amazon-linux-extras install python3
pip install numpy boto

# Get the classifier code from the shared URL (the same bucket on this
    case)
curl https://s3-eu-west-1.amazonaws.com/nasa-experiment/classifier.py -o
    /tmp/classifier.py
chmod +x /tmp/classifier.py
/tmp/classifier.py

# Computation and upload finished: the instance will terminate itself
shutdown -h now
```

The main steps in the simulation are as follows:

- Obtain the cloud score for a month (day by day) from the NASA API.
- Given that data, calculate the average cloud score for the month and the most typical day of that month.
- Upload the result to an AWS S3 object.
- Return a success message to the user.

---

[2] Note that this code is greatly simplified (for illustrative purposes) and is subject to optimization (such as multiprocess requests using arguments instead of constants).

[3] The installation steps can be found at https://docs.aws.amazon.com/cli/latest/userguide/installing.html.

[4] Note, as previously mentioned, that the code of the classifier must be publicly available from the URL that the script curls.

Finally, the code of *classificator.py* is as follows:

```python
import json
import numpy
import os
import requests
import tempfile
import boto
import boto.s3
from calendar import monthrange
from datetime import datetime
from boto.s3.key import Key

# Constants: Keys and secrets
NASA_API_KEY = "NASA_API_KEY"
AWS_ACCESS_KEY_ID = 'XXX'
AWS_SECRET_ACCESS_KEY = 'YYY'
S3_BUCKET = 'MY_EXPERIMENT_BUCKET'

# Constants: Date YYYY-MM
DATE = "2016-07"

# Constants: Longitude and Latitude
LON = "-7.8987052"
LAT = "42.3384207"


class S3(object):

    def __init__(self):
        self._conn = boto.connect_s3(AWS_ACCESS_KEY_ID,
            AWS_SECRET_ACCESS_KEY)
        self._bucket = self._conn.create_bucket(
            S3_BUCKET, location=boto.s3.connection.Location.DEFAULT)

    def upload(self, source_file, target_file):
        k = Key(self._bucket)
        k.key = target_file
        k.set_contents_from_filename(source_file)


class Month(object):

    def __init__(self, longitude, latitude, date, api_key):
```

6.3 Starting an Experiment in the Cloud

```python
        self._longitude = longitude
        self._latitude = latitude
        self._date = date
        self._api_key = api_key

        self._baseurl = "https://api.nasa.gov/planetary/earth/imagery/"
        self._req = requests

        # Data from month
        self._monthdata = {}

    def _getData(self):
        year, month = self._date.split('-')
        mrange = monthrange(int(year), int(month))

        print("> Receiving data from API...")

        for day in range(1, mrange[1] + 1):
            day_date = "{}-{}".format(self._date, day)
            day_url = "{base}?lon={lon}&lat={lat}\
                    &date={date}&cloud_score=True\
                    &api_key={api_key}".format(base=self._baseurl,
                    lon=self._longitude, lat=self._latitude,
                    date=day_date, api_key=self._api_key)

            # An optimization could be to make less requests to the API (
                closest day from the API)
            # Another optimization is to use multiprocessing pool for
                parallel
            # requests
            day_data = self._req.get(day_url)
            day_data_json = day_data.json()
            response_date = datetime.strptime(
                day_data_json['date'], '%Y-%m-%dT%H:%M:%S')
            response_simple_date = "{}-{}-{}".format(
                response_date.year, response_date.month, response_date.
                    day)

            # Add data to dict only if not there already
            if response_simple_date not in self._monthdata:
                self._monthdata[response_simple_date] = day_data_json

    def _getAverageDay(self):
        self._getAverageCloudScore()
```

```python
            closest_average_day = ''
            average_distance = 1

            for day in self._monthdata:
                day_avg_distance = numpy.absolute(
                    self._monthdata[day]['cloud_score'] - self.
                    _average_cloud_score)
                if (day_avg_distance < average_distance):
                    # Mark this day as the closes to the average value
                    average_distance = day_avg_distance
                    closest_average_day = day

            self._closest_average_day = closest_average_day
            self._writeTempFile()

        def _getAverageCloudScore(self):
            cloud_scores = []
            for day in self._monthdata:
                cloud_scores.append(self._monthdata[day]['cloud_score'])

            self._average_cloud_score = numpy.mean(cloud_scores)

        def _writeTempFile(self):
            new_file, filename = tempfile.mkstemp()
            print("> Writing local temp file: %s" % filename)
            os.write(new_file, json.dumps(
                self._monthdata[self._closest_average_day]))
            os.close(new_file)

        def _writeOutput(self):
            print(" ")
            print("> Month Average Cloud Score for %s: %s" % (DATE, self.
                _average_cloud_score))
            print("> Average Day Date (YYYY-MM-DD): %s (avg.: %s)" %
                (self._closest_average_day, self._monthdata[self.
                    _closest_average_day]['cloud_score']))

        def process(self):
            self._getData()
            self._getAverageDay()
            self._writeOutput()
            self._uploadS3()
```

```
if __name__ == "__main__":

    month = Month(LON, LAT, DATE, NASA_API_KEY)

    # Process data for month
    month.process()
```

References

1. Mell P, Grance T (2011) SP 800-145. The NIST definition of cloud computing. National Institute of Standards & Technology, Gaithersburg
2. Reuther A, Byun C, Arcand W, Bestor D, Bergeron B, Hubbell M, Jones M, Michaleas P, Prout A, Rosa A, Kepner J (2018) Scalable system scheduling for HPC and big data. J Parall Distr Com 111:76–92
3. Wagstaff K (2012) What, exactly, is a supercomputer? Time. http://techland.time.com/2012/06/19/what-exactly-is-a-supercomputer/. Accessed 9 Aug 2018
4. Andrikopoulos V, Binz T, Leymann F, et al (2013) How to adapt applications for the cloud environment. Computing 95:493. https://doi.org/10.1007/s00607-012-0248-2
5. Eizinger T (2017) API design in distributed systems: a comparison between GraphQL and REST. Master Thesis, University of Applied Sciences Technikum Wien
6. NASA (2019) Earthdata Developer Portal. https://earthdata.nasa.gov/collaborate/open-data-services-and-software/api. Accessed 5 Nov 2019
7. Google (2019) Google API explorer. https://developers.google.com/apis-explorer/. Accessed 20 May 2019
8. Microsoft (2019) Azure REST API reference. https://docs.microsoft.com/en-us/rest/api/azure/. Accessed 22 May 2019
9. Amazon (2019) AWS documentation. https://docs.aws.amazon.com/#lang/en_us. Accessed 10 Apr 2019
10. Google (2019) Google cloud APIs. https://cloud.google.com/apis/docs/overview. Accessed 20 May 2019
11. Amazon (2019) AWS SDK. https://aws.amazon.com/tools/#SDKs. Accessed 10 Apr 2019
12. Google (2019) Cloud SDK. https://cloud.google.com/sdk/. Accessed 20 May 2019
13. Microsoft (2019) Azure tools. https://azure.microsoft.com/en-us/tools/. Accessed 22 May 2019
14. Vijayasarathy L, Butler C (2016) Choice of software development methodologies: do organizational, project, and team characteristics matter? IEEE Softw 33(5):86–94. https://doi.org/10.1109/MS.2015.26
15. Amazon (2019) AWS VPC documentation. https://docs.aws.amazon.com/vpc/index.html. Accessed 10 Apr 2019

Chapter 7
Serverless Experiments in the Cloud

7.1 Sasha Revisits the Experiment

After spending time running a small experiment on the cloud over EC2, i.e., the classic approach, Sasha realized that the setup was complex. Serverless computing provides a promising new solution for this problem. Several examples of the application of serverless computing in the field of scientific research exist, as shown in Sect. 3.6, and Sasha decides to explore whether it could simplify the code, the setup, and the way the team runs simulations. Let us see what serverless computing means and how it can make cloud computing easier to implement.

7.2 Thinking Serverless

Function as a Service (FaaS), or serverless (computing) [1] is a relatively new model within the cloud paradigm in which software runs in a highly abstracted and constrained environment and the developer has a low level of control in exchange for an automated and hands-off infrastructure. Serverless computing platforms vary in terms of the level of abstraction from those that allow developers to have some control, such as Google's App Engine, where machine type can be selected and third party packages can be installed, to FaaS such as AWS Lambda, where the runtime environment is entirely defined by AWS and the developer has to adjust their code to the restricted environment.

Designing (or re-designing) a solution for FaaS [2] requires re-thinking some aspects such as how to trigger the function and how to stream data; therefore, serverless computing is not for every one and every problem but is highly advantageous for certain types of implementations. If Sasha really needs control of their resources (e.g., disk IOPS) or their project has specific or customized requirements (e.g., custom libraries), serverless computing would not be the best solution. However, if Sasha's software is sufficiently generic (e.g., using a major language with standard numerical modules), solutions within the serverless spectrum might be optimal.

7.2.1 Pros and Cons

Pros
- **Simplified workflow:** This benefit not only defines but also justifies the use of serverless computing over any other cloud service. The cloud provider offers the required expertise to run the infrastructure, so the developer is hands-off and can focus their efforts on designing and implementing software.
- **Better control over costs:** Serverless computing (usually) follows a pay-as-you-go scheme, where the provider charges only for the exact used resources at a finer-grain level than that of any other cloud service.

Cons
- **Challenging for large HPC problems [3, 4]:** Allocating a large number of resources still requires planning with the cloud provider; otherwise, Sasha might experience a situation where, for example, the CPU power required by their application is not available.
- **Higher level of abstraction corresponds to a lower level of control:** A good example is that Sasha's software could suffer underperformance if the wrong resources are used. Remember that Sasha does not control many of the resources: the Cloud provider allocates them (and the resources might not be a good fit for their problem).

- **Specific code:** In some cases, Sasha's application could require customized code that is available only on certain providers, that is, non-standard code and some level of vendor lock-in. Moreover, Sasha should evaluate if resources are available to refactor their code to run within a serverless environment.

7.3 Example: Month Classifier, Serverless Version

After revisiting the month classifier experiment (described in the previous chapter) and reviewing the current state of the art and trends in cloud computing, Sasha thinks that their month classifier is a good candidate for a serverless implementation for the following reasons:

- The researchers will not need to consider the requirements and dependencies such as EC2 instances or VPC.
- The code is standard (Python 3) and relatively simple.
- The additional costs related to the infrastructure will be reduced using on-demand computing.

In this case, Sasha has received some funding for GCP, so the FaaS solution is selected: Google Cloud Functions [5]. The specific details are as follows:

- **Cloud provider:** In this case, let us assume that it is GCP.
- **Technologies:**
  - Programming language and libraries: Python 3 and NumPy.
  - Third-party APIs: NASA's Earth data API.
  - Cloud technologies/services: Google Cloud Functions and Google Cloud Storage (GCS).

  A very high-level architecture and workflow is shown in Fig. 7.1.
- **Pre-requirements:**
  - Valid credentials (service account) that have permission to manage both cloud functions and cloud storage are required. The easiest way to obtain credentials is to use the Google Cloud SDK [6] and authenticate the current session with the following command to provide a link to open and validate the session:

  ```
  $ gcloud auth login
  ```

  - A GCS bucket must be created, which can be done by running the following command:

  ```
  $ gcloud mb gs://MY_EXPERIMENT_BUCKET
  ```

Note that this code is a simple approach and is not intended for production; additionally, this is a sync call (long-running sync HTTP requests are usually not a good idea) used as an example starting point. This code is a variation of the

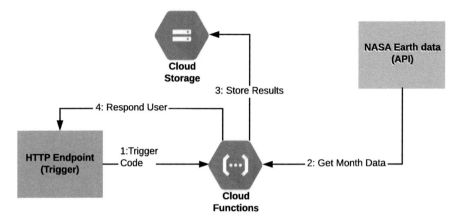

Fig. 7.1 High-level serverless computing architecture and workflow for the cloudiness month classifier experiment

that presented in Sect. 6.3.1 but adapted to Google Cloud Functions. The main differences are the following:

- The requirements are resolved on the server side, and Google takes care of the resolution (out of a *requirements.txt* file).
- The output and logs are gathered by the cloud platform and presented on a UI (Google Cloud Logs/Stackdriver in this case).
- All the authentication and storage that was AWS-specific has been replaced by the GCS counterpart.

Despite these differences, the main logic of the simulation remains the same:

- The code is triggered, in this case, by the HTTP endpoint created by Google Cloud Functions for this experiment.
- Obtain the cloud score for a month (day by day) from the NASA API.
- Given that data, calculate the average cloud score for the month and the most typical day of that month (closest to the average).
- Upload the result to a Google Cloud Storage object.
- Return a success message to the user.

Two files are required to run this program: the requirements file (requirements.txt, used to resolve dependencies) and the simulation code itself (main.py).

The content of the file *requirements.txt* is as follows:

```
numpy
google-cloud-storage
```

The content of the file *main.py* is as follows:

```
import json
```

7.3 Example: Month Classifier, Serverless Version

```python
import numpy
import os
import requests
from calendar import monthrange
from datetime import datetime
from google.cloud import storage

# Constants: Keys and secrets
NASA_API_KEY = "NASA_API_KEY"
AWS_ACCESS_KEY_ID = 'XXX'
AWS_SECRET_ACCESS_KEY = 'YYY'
GCS_BUCKET = 'MY EXPERIMENT BUCKET'

# Constants: Date YYYY-MM
DATE = "2016-07"

# Constants: Longitude and Latitude
LON = "-7.8987052"
LAT = "42.3384207"


class Month(object):

    def __init__(self, longitude, latitude, date, api_key):
        self._longitude = longitude
        self._latitude = latitude
        self._date = date
        self._api_key = api_key

        self._baseurl = "https://api.nasa.gov/planetary/earth/imagery/"
        self._req = requests

        # Data from month
        self._monthdata = {}

    def _getData(self):
        year, month = self._date.split('-')
        mrange = monthrange(int(year), int(month))

        print("> Receiving data from API...")

        for day in range(1, mrange[1] + 1):
            day_date = "{}-{}".format(self._date, day)
```

```python
            day_url = "{base}?lon={lon}&lat={lat}&date={date}&
                cloud_score=True&api_key={api_key}"
                    .format(base=self._baseurl, lon=self._longitude,
                        lat=self._latitude,
                            date=day_date, api_key=self._api_key)

            # An optimization could be to make less requests to the API (
                closest day from the API)
            # Another optimization is to use multiprocessing pool for
                parallel
            # requests
            day_data = self._req.get(day_url)
            day_data_json = day_data.json()
            response_date = datetime.strptime(
                day_data_json['date'], '%Y-%m-%dT%H:%M:%S')
            response_simple_date = "{}-{}-{}".format(
                response_date.year, response_date.month, response_date.
                    day)

            # Add data to dict only if not there already
            if response_simple_date not in self._monthdata:
                self._monthdata[response_simple_date] = day_data_json

    def _getAverageDay(self):
        self._getAverageCloudScore()
        closest_average_day = ''
        average_distance = 1

        for day in self._monthdata:
            day_avg_distance = numpy.absolute(
                self._monthdata[day]['cloud_score'] - self.
                    _average_cloud_score)
            if (day_avg_distance < average_distance):
                # Mark this day as the closes to the average value
                average_distance = day_avg_distance
                closest_average_day = day

        self._closest_average_day = closest_average_day

    def _getAverageCloudScore(self):
        cloud_scores = []
        for day in self._monthdata:
            cloud_scores.append(self._monthdata[day]['cloud_score'])
```

7.3 Example: Month Classifier, Serverless Version

```python
        self._average_cloud_score = numpy.mean(cloud_scores)

    def _writeResultFile(self, filename, file_content):
        # Write Result file to GCS Bucket
        client = storage.Client()
        bucket = client.get_bucket(GCS_BUCKET)
        blob = bucket.blob(filename)
        blob.upload_from_string(str(file_content))

    def _writeOutput(self):
        # This will be shown on the Function logs output
        print(" ")
        print("> Month Average Cloud Score for %s: %s" % (DATE, self.
            _average_cloud_score))
        print("> Average Day Date (YYYY-MM-DD): %s (avg.: %s)" % (
            self._closest_average_day, self._monthdata[self.
            _closest_average_day]['cloud_score']))

    def process(self):
        self._getData()
        self._getAverageDay()
        self._writeOutput()
        # Write month average to storage
        self._writeResultFile("%s_%s" % (DATE, "month_average"), self.
            _average_cloud_score)
        # Write typical data for a day a (this is, average day) to storage
        self._writeResultFile("%s_%s" % (DATE, "day_average"), self.
            _closest_average_day)

def classifier(request):

    month = Month(LON, LAT, DATE, NASA_API_KEY)

    # Process data for month
    month.process()

    return f"Computation completed! Results available on the GCS Bucket."
```

Once the code is ready, it can be deployed to a cloud function with the following simple command:

```
$ gcloud functions deploy start_simulation --runtime python37 --trigger
    -http
```

This command will return information about our function, such as the endpoint (httpsTrigger url), which will be used to trigger the cloud function. For our example, the function is as follows: https://us-central1-OUR-EXPERIMENT.cloudfunctions.net/simulation. Then, the command to start the simulation is as follows:

```
$ curl "https://us−central1−OUR−EXPERIMENT.cloudfunctions.net/simulation"
```

Once the simulation is complete, it can be removed with the gcloud command:

```
$ gcloud functions delete simulation
```

7.4 Example: Serverless Sensor Data Collector

After the success running the month classifier over Google Cloud Functions, Sasha's team have been considering taking advantage of their new knowledge to solve another problem they have faced for a while: NASA API's data is great but is not exactly what they need. Sasha's team is more interested in the cloud coverage for a specific region and have been deploying highly sensitive sensors that can send real-time data. Thus, the researcher needs an easy way to collect these data.

Given that the team now knows how to run serverless code, Sasha suggests a simple function that, given a sensor that sends data, writes the result to a file (for now, they will probably start with using a key/value store, such as Datastore) so that they can read the results directly from their GCS bucket. The workflow for this task is shown in Fig. 7.2.

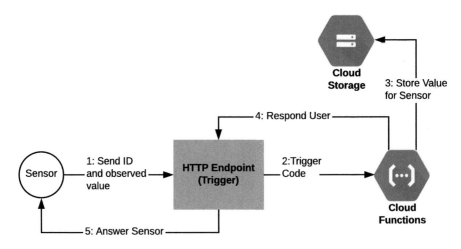

Fig. 7.2 High-level serverless architecture and workflow for a sensor data collector

7.4 Example: Serverless Sensor Data Collector

The implementation is as follows[1]:
The content of the file *requirements.txt* is the following:

```
google-cloud-storage
```

The content of the file *main.py* is the following:

```python
import time
import json
from google.cloud import storage

# Constants
GCS_BUCKET = 'MY_EXPERIMENT_BUCKET'

def writeSensorData(sensor_id, sensor_value):
    # Write Result file to GCS Bucket
    timestamp = int(time.time())
    filename = "sensor_%s_%s"%(sensor_id, timestamp)
    client = storage.Client()
    bucket = client.get_bucket(GCS_BUCKET)
    blob = bucket.blob(filename)
    blob.upload_from_string(str(sensor_value))

def sensor(request):
    # sensor_id and sensor_value are required arguments
    request_json = request.get_json()
    if request.args and 'sensor_id' in request.args and 'sensor_value' in request.args:
        args = request.args.to_dict()
        writeSensorData(args['sensor_id'], args['sensor_value'])
        return f'OK'

    return f'Missing Arguments!'
```

The code is deployed via the following command:

```
$ gcloud functions deploy sensor --runtime python37 --trigger-http
```

Now, the sensors simply need to call a URL to start reporting their data to the serverless microservice:

```
curl "https:///us-central1-OUR-EXPERIMENT.cloudfunctions.net/sensor?sensor_id=<SENSOR_ID>&sensor_value=<SENSOR_VALUE>"
```

---

[1] Note that there is no security or authentication in this implementation; it is simply a proof of concept: anyone with the URL can make a GET request and insert data in the GCS bucket.

References

1. Baldini I et al (2017) Serverless computing: current trends and open problems. Research advances in cloud computing. Springer, Singapore, pp 1–20. https://doi.org/10.1007/978-981-10-5026-8_1
2. Hong S (2018) Go Serverless: securing cloud via serverless design patterns. In: 10th USENIX workshop on hot topics in cloud computing (HotCloud 18), Boston
3. Kleppmann M (2017) Designing data-intensive applications, Chapter 8. O'Reilly, Sebastopol
4. Montes D et al (2017) Views on the potential of Cloud computing and IaaS/HPCaaS for meteorology and climatology. In: EMS Annual Meeting Abstracts (EMS2017), Vienna, vol 14, p 639
5. Google (2019) Google cloud functions. https://cloud.google.com/functions/docs/concepts/python-runtime. Accessed 10 May 2019
6. Google (2019) Cloud SDK. https://cloud.google.com/sdk/. Accessed 10 May 2019

Chapter 8
Ethical and Legal Considerations of Cloud Computing

For the purpose of this book, addressing ethical and legal issues is more challenging than initially thought. First, Sasha must realize that both scientific research and IT act within their own realms. When conducting research, depending on the project, Sasha might need their work to be approved by a commission of ethics. Moreover, Sasha must follow the scientific method to ensure scientific reproducibility and make the work useful to others.

Ethics in information technology is not very different from ethics in other fields. Given that Sasha probably does not develop most of their software and hardware, Sasha must consider what software they are using and how it has been designed. This factor was noted decades ago by Richard M. Stallman when he initiated the Free Software movement, and it is also part of the activity developed by bodies such as the Free Software Foundation and the Electronic Frontier Foundation.

After the popularization of the Internet, data protection added a new dimension. In some ways, the use of cloud computing involves the same legal issues as working with many other processes and services on the Internet. For example, Sasha should know whether the SaaS being used is under a license of the style of the Affero GPL (AGPL) [1] (specifically designed for the cloud), which would provide a guarantee that the software run in the cloud is free, by granting access to its code.

Companies selling cloud computing services must be the first to provide maximum transparency and compliance with the highest standards. Users of these services include National Agencies, for example, the military, which require the strongest possible security levels. Fortunately, companies providing these services must comply with strong ISO quality standards that ensure high levels of responsibility.

Almost 10 years ago, there was an initiative to develop a "Cloud Computing Manifesto" in a collaborative manner based on free tools and with the participation of industry. The manifesto was a step beyond the "Cloud Computing Bill of Rights"; however, the main players eventually decided not to sign the manifesto, and the idea died in 2012. Drafts of the manifesto were highly technical, focusing on the development of the business model of cloud computing; however, they also included the notion of interoperability among "Clouds." Recent work [2] classifies the ethical issues in cloud computing according to four categories: privacy and security, compliance, performance ethics and environmental impacts (already discussed in Sect. 3.5). The first category is the typical worry about data privacy, the second concerns the correct level of standards. Performance ethics refer to the delivery of the service that has been contracted. In summary, these three concerns are related to general "business ethics" and are applicable independent of whether the cloud is being used or not. The implications and problems that cloud computing poses in this field are acknowledged by the existence of recent projects such as SafeCloud [3], which is funded by the European Union.

At this point, Sasha could be using cloud computing resources for research at any level of implementation or any combination presented in the previous chapters. One of the main concerns raised in cloud computing is related to data. Giving others custody of your data always raises legal and ethical issues that must be taken into account. Ethical issues related to potential data mining of your stored data might not be your main concern if you are storing personal photos or emails but could be worrying if you store medical records from patients in clinical trials. For example, to address this latter problem, AWS implements measures to protect medical data and comply with laws such as the U.S.A. HIPAA Act [4]. However, in some cases, data storage can raise complicated issues, for example, using data handed over by a third party such as a private company. In such cases, the number of copies of the data and the storage location could have been settled by a previous contract that could eventually come into conflict with the contract that you signed with a given cloud computing provider. A good example is the problem raised after Google has got access to medical records of Americans [5].

Furthermore, in scientific research, all the details related to the software being used must be known, including versions and license (which should be free [6]), because these details are mandatory to ensure scientific reproducibility of your results and should be made available in publications. Indeed, there is a need for increasing scientific computational reproducibility; therefore, it is necessary to ensure that cloud computing is in compliance. This consideration is also useful to avoid lock-in situations. For example, you may eventually want to move your operations and data from one cloud provider to another. From a technical perspective, this process will be much easier given more information about your current infrastructure.

In summary, when using cloud computing, Sasha should be aware of the lack of full control on what is shared. For example, a given cloud provider could be forced by governmental bodies to share Sasha's data because of legal concerns. Plenty of examples of this scenario exist, for example, the U.K.-U.S. Cloud Act signed in October 2019 [7]. This scenario applies to only a minor portion of cases of scientific research, and some kind of illegal activity would need to be spotted. However, the main point is that by using cloud computing, Sasha cedes power to make the final decision of whether to share data (although additional possibilities, such as the use of encrypted cloud storage, exist). Moreover, Sasha should be aware that data stored in the cloud are likely not protected or under the same laws and regulations as those in their country or usual legal domain, as the infrastructure providing the cloud service is located in several countries and the company may be based in another country. Therefore, the data may exist under potentially very different legal conditions. The exact legal conditions partially depend on the contract a user signs with a cloud provider—which can vary substantially—and how the cloud provider applies the contract [8]. In other cases, local regulations such as those of the European Union could be incompatible with regulations of other regions. You should be aware that you might have to specifically choose that your research is conducted and your data are stored on data centers geographically located in the same region as your institution or funding body. This is not only because of the laws that apply to data but the right of the members of Sasha's team to access them and the resources, sometimes only available for certain people that have got security clearance or has the right passport.

References

1. GNU (2007) GNU Affero General Public License. https://www.gnu.org/licenses/agpl-3.0.html. Accessed 10 Feb 2019
2. Faragardi H (2017) Ethical considerations in cloud computing systems. Proceedings 1:166. https://doi.org/10.3390/IS4SI-2017-04016

3. EASA (2016) SafeClouds.eu Data-driven research addressing aviation safety intelligence. https://ec.europa.eu/inea/en/horizon-2020/projects/h2020-transport/aviation/safeclouds.eu. Accessed 05 Apr 2019
4. Amazon (2019) HIPAA compliance. https://aws.amazon.com/compliance/hipaa-compliance/. Accessed 06 Apr 2019
5. Copeland R (2019) Google's 'Project Nightingale' gathers personal health data on millions of Americans. Wall Street J. https://www.wsj.com/articles/google-s-secret-project-nightingale-gathers-personal-health-data-on-millions-of-americans-11573496790
6. Añel JA (2011) The importance of reviewing the code. Commun ACM 54(5):40–41. https://doi.org/10.1145/1941487.1941502
7. Department of Justice of the United States (2019) Cloud act resources. https://www.justice.gov/dag/cloudact. Accessed 15 Apr 2019
8. Srinivasan S (2014) Cloud computing basics. Springer, New York

Chapter 9
You Are Outdated, We Are Already Updating This Book

I'm sorry, are your from the past? (Roy, The IT Crowd)

That innovation is happening all the time and that the pace of innovation has not been equal over time is clear. Throughout the past century, the pace of innovation has increased dramatically. In fields such as computer science, innovation is a major factor. Cloud computing has become a dominant technology in recent years, driving innovation and business profits. As cloud computing develops and adapts in response to people's desires and needs to manage data and devices in ways that go beyond purely technological advances, substantial room for improvement exists in the next few years.

Clearly, the strategies of some providers of cloud computing services will include offering more complete solutions and enhancing the regular services used for in-house computing systems in the cloud. For example, Google has recently announced that it is making several blockchain technologies available in its Cloud services. Additionally, the bid for the integration of AI as a side service to offer jointly with cloud computing for different purposes (e.g., data mining) is now clear. This kind of service has great potential, as in some cases, you only need to allow access to data already stored in the cloud without any additional work on your part to obtain further analysis, insights or recommendations that might eventually lead you to new ideas, results or improvement of management or other practices.

As the use of cloud services continues to grow, the business of consultancy to adapt existing working environments (laboratories, companies, etc.) to cloud computing and to assess and perform migrations is expected to grow dramatically in the future.

Another emerging field is quantum computing, a technology that is clearly not suitable for desktop users. It is almost inconceivable that a quantum computer could be operated outside a laboratory under very specific conditions to maintain the lowest temperatures known. Therefore, quantum computing is likely to be operated as a cloud computing service (IBM already offers something similar through their "IBM Q Experience" (https://www.ibm.com/quantum-computing/technology/experience/)), where the only thing that changes is the hardware used to provide service. Therefore, in the future, any attempt to advance quantum computing will follow the cloud computing paradigm. Although this kind of statement in computer sciences can become obsolete in only a few years, we can state that no one expects to sell a physical quantum computer in the near future.

The broader adoption of cloud computing services is expected to drive investment in better Internet connections around the world. Some forecasts state that by the end of the next decade, we will not be talking about connectivity and the use of the cloud, but that it will be the only technology and everything will be interconnected, including technologies such as the IoT. Moreover, a growing research line related to networks in cloud computing is the use of such technology to improve cybersecurity, for example, to increase resilience to DDoS attacks.

As a scientist, Sasha usually applies for and has to manage funding. Sasha could be thinking about the next computer that must be purchased for the next research project. If Sasha is not thinking about a quasi-portable device such as a laptop or a small box, then they are stuck in the past. Other than privacy and constant access to the most relevant data needed for a few days or everyday, there is no point in storing data on a large machine. If the word "workstation" is familiar to you and is how you usually work, check Wikipedia, which speaks about this kind of computer mostly in past tense.

You might be worried about what happens if you travel to a place with no Internet connection and cannot access your data. This situation is not going to disappear entirely, but in some regions of the world, this scenario is comparable to worrying about not having a power connection. Furthermore, there are places where your chances of having Internet access are greater than those of finding a power switch.

Glossary

Agile Agile (Software development) A term that comprises different methodologies, practices, and frameworks based on cross-functional self-organized teams that aim to deliver products and services in a continuously improving, incremental and fast manner. Examples of Agile approaches are Scrum, Kanban, and DevOps.
Alexa A device developed and sold by Amazon that works as a personal assistant through connection to the cloud.
Apache Short form of *Apache HTTP Server*. One of the most popular web server software.
APIs Application programming interfaces (APIs) are (very) high-level abstractions that provide a standard way of communicating with software or infrastructure.
AWS Short form of *Amazon Web Services*, Amazon's cloud computing platform.
AWS Lambda AWS FaaS implementation.
Azure Microsoft's cloud computing platform.
Big Data Computer science field that processes and analyses big sets of data for different purposes, such as extraction of meaningful information or forecasting.
Blockchain Series of technologies that power the aggregation (append) of a set of records connected via cryptographic methods. Common uses for blockchain are electronic wallets and smart contracts.
BOINC Free-software middleware for grid computing, implemented by the University of California (Berkeley) and widely used for volunteer computing.
Bucket Highest level in the hierarchy of cloud storage systems (such as S3 or GCS) that acts as a container for the objects of the file system.
Chatbot Software that performs conversations.
Citizen science Science performed by citizens, usually contributing to basic parts of the research process and in collaboration or under the supervision of professional scientists.

Cloud enabled Software or system that initially was not created to run on the cloud but that, with some modifications, can run in such environment. Usually, cloud-enabled software can take advantage of all the features that cloud computing offers.

Cloud native Software or system that was designed to run to take advantage of all the features that the cloud computing environment offers. Usually, cloud-native software cannot run outside the cloud or will lose many of its features outside of this environment.

CNCF Cloud Native Computing Foundation, part of the Linux Foundation. Created to promote the concept of cloud native and facilitate the creation and growth of an ecosystem.

Container Virtual runtime environment that runs on top of a single operating system (OS) kernel and emulates an operating system rather than the underlying hardware.

Cores Independent processing units that can be grouped into several integrated circuits.

Cpanel Online Linux-based web hosting control panel.

Data mining Processes and methodologies aimed towards discovering and extracting patterns from data.

DDoS Distributed denial of service. Attack that causes a denial of service (DOS) over a system or service.

Docker Most used PaaS to run containers.

EC2 AWS implementation to run contextualized virtual machines (instances) in the cloud.

ERP Integrated suite for business management.

FaaS A way of using cloud computing associated to serverless that requires only the implementation of functions without consideration of the underlying infrastructure.

Free Software Free software refers to software that complies with the definition of the free-software foundation (https://www.gnu.org/philosophy/free-sw.en.html) and implies freedom to run the software, to access the code and modify it, to redistribute it, and to redistribute modified versions of the program.

GCP Short form of *Google Cloud Platform*, Google's cloud computing platform.

Git A distributed version of control system widely used in software development.

GNU Operative system that is free software (free as in "freedom"). GNU is a recursive acronym for "GNU's not UNIX." The most extensive versions run with the Linux kernel, though other versions are available.

Google App Engine Also known as GAE. Highly abstracted and sandboxed PaaS implementation from GCP.

Google Cloud Functions GCP FaaS implementation.

GraphQL Facebook open-source data query and manipulation language for APIs. Alternative to REST.

Hadoop Apache's big data platform based on the massive distributed MapReduce programming model.

InfiniBand Networking communications standard used in HPC.

Glossary

Instance Virtual machine with context (CPU definition, memory size, disk image, etc.) running in a cloud computing environment. In the context of this book, the terms *virtual machine* and *instance* are interchangeable.

IOPS Performance measurement that indicates input/output operations per second.

IRC A text chat protocol created by the end of the 1980s that became extremely popular during the 1990s.

JSON JavaScript object notation is an open-standard file format that uses human-readable text to transmit data objects.

Kernel Computer program, the core part of the operating system of a computer.

Kubernetes Free-software (container) orchestration system created by Google.

LaTeX A document preparation system that is commonly used in scientific environments.

Linux Extremely popular kernel for operating systems.

Machine learning Cross-discipline field (with strong base in computer sciences and statistics) that aims to extract and infer patterns in pools of (a priori) non-structured data.

Microservice Architecture composed of decoupled, small, and independent services.

MPI Standard for parallelization of tasks, usually in HPC environments.

NAS Storage attached and shared over the network.

NumPy Numerical library (and extensions) to Python.

Observability Attribute of systems and services that are observable (historical data or forecasting) through metrics and logs.

PHP High-level programming language, very popular for web development.

POSIX A series of standards with the goal of maintaining compatibility across different operating systems.

Private Cloud Cloud model where the infrastructure can be managed directly by the IT department of the client or by the vendor, according to needs.

Private network Network that is private to an organization or infrastructure (that is, not publicly accessible).

Public Cloud Cloud model where the servers in a data center can be shared by several users according to needs. Resources are shared and publicly available.

PUE Power usage effectiveness is a method (ratio) to calculate how efficient a data center is in terms of energy consumption.

Python High-level programming language, very popular in many environments but especially in science, data science, and systems development.

Region A cloud region is the way that cloud providers represent the geographical location of their data centers. Regions are divided in zones.

Research Credits Waivers and credit given by cloud computing providers to researchers.

REST Representational state transfer architecture. Used to develop (RESTful) APIs.

S3 AWS cloud storage implementation.

SDK Related software developer tools packaged together.

SEO Techniques used to make websites rank higher on search engines like Google, Bing, and Yahoo Search.

Serverless See FaaS.

Spark Apache Spark is a cluster computing framework built after Hadoop. Its main architecture is based on resilient distributed dataset (RDD), an in-memory data structure that allows fast computation.

SSD Solid state disk, usually flash memory that is used to achieve low latency and high performance.

SSL See TLS.

TLS Transport layer security (that deprecates SSL) is a series of cryptographic protocols that aim to provide secure communications over a network.

Virtual machine See instance.

Virtualization In the context of this book, this term means creating or running an instance or container that runs as a real operating system.

VPC A virtual private cloud is a cloud environment that is isolated from others (its network is not connected to a public network) and is usually only reachable via a VPN.

VPN A virtual private network is a concept that indicates that a private network is extended over a public network, that is, two or more private networks (not reachable over public networks) can see each other and use public networks as their communication channel (usually encrypted and secure).

VPS A virtual private server is a virtual machine sold by a provider. VPSs live in a lower level of abstraction than cloud instances and are sold as pre-installed (as-it-is) systems (providers do not allow self-service for VPS).

Zero download The scenario in which data are maintained in external servers (the cloud) and computational tasks are performed there instead of maintaining big datasets in-house or local systems to perform computations on them.

Index

A
Affero GPL, 74
Agile, 43
AI, 28, 78
Alexa, 45
Amazon, 33
Apache, 23
Application programming interface (API), 24, 38, 49, 50, 53–55, 65
AWS, ix, 19, 22, 24, 26, 30, 33, 36, 45, 51, 55, 57, 66, 74
AWS Cloudwatch, 24
AWS Lambda, 22, 30, 64
Azure, 19, 21, 26, 27, 33, 51

B
Bandwidth, 11
Big Data, 2
Bitbucket, 8, 44
Blockchain, 78
BOINC, 28

C
CDC, 29
CERN, 30
Citizen science, 20
CloudatHome, 20
Cloud data mining, 3, 35
Cloud enabled, 53
Cloud native, 24, 53
Cloud Native Computing Foundation (CNCF), 25, 26
Container, 25

D
Data mining, 3, 78
DDoS, 78
Docker, 25, 54
Dropbox, ix, 22

E
EC2, 22, 51, 54, 56, 63, 65
Environmental impact, 26
ERP, 9
ESA, 30

F
FDA, 29
FMI, 28
Free software, 18–20, 24, 26, 30, 44, 74
Function as a Service (FaaS), 22, 64, 65

G
GCP, 19, 26, 30, 51, 65
GCP Stackdriver, 24
Git, 7, 44, 55
GitHub, 8, 42, 44
GitLab, 8, 44
Gmail, 22
Google, ix, 19, 26, 36, 42, 49, 51, 75, 78
Google App Engine, 64
Google Cloud Functions, 22, 65, 66
Google Cloud Storage (GCS), 8, 65, 66, 70
Google Dataproc, 22
Google Docs, 42
Google Hangouts, 42

© Springer Nature Switzerland AG 2020
J. A. Añel et al., *Cloud and Serverless Computing for Scientists*,
https://doi.org/10.1007/978-3-030-41784-0

Graphana, 24
Graphite, 24
G Suite, 42

H
Hadoop, 22, 54
High-performance computing (HPC), 23, 48, 49, 54
HPCaaS, 1, 5, 23, 28, 29, 34
Hybrid cloud, 21, 40

I
IBM, 19
IDE, 45
Infiniband, 54
Infrastructure as a Service (IaaS), 5, 11, 22, 27, 28
Internet of Things (IoT), 22, 78
IOPS, 64
IRC, 42
ISO, 74

J
Jira, 42–44
JSON, 51

K
Kernel, 25, 44
Kubernetes, 26, 54

L
Latency, 54
LaTeX, 43
License, 74

M
Machine learning, 2
Met Office, 28
Microservices, 26
Microsoft, ix, 19, 27, 33, 42, 44
MPI, 54
Multicloud, 40

N
NAS, 7
NASA, 20, 30, 49, 55, 65
NCAR, 28
Network, 54
NOAA, 28

NREL, 28
Numpy, 56, 65

O
Office 365, 42
OpenFoam, 30
OpenNebula, 20
OpenStack, 20
Overleaf, 43

P
PHP, 23
Platform as a Service (PaaS), 11, 22, 25, 26, 28, 54
POSIX, 49
Power usage, 27
Privacy, 74
Private cloud, 21
Processor, 42
Proprietary software, 18
Public cloud, 20, 51
Python, 51, 56, 65

Q
Quantum computing, 78

R
Rackspace, 19, 20
RAM, 11
Red Hat, 19
Region, 51, 54, 75
Reproducibility, 26, 73, 75

S
S3, 24, 56, 57
SDK, 51, 65
SEO, 10
Slack, 42
Software as a Service (SaaS), 11, 22, 27, 28, 41, 74
Spark, 22
SSH, 56
SSL, 11
Storage, 55
Sustainability, 33

T
TLS, 11
Trello, 43

U
UI, 50, 54, 56, 66

V
Virtual private cloud (VPC), 56, 65
VPN, 55

W
Workstation, 48, 78

Z
Zero download, 3